ADVANCES IN DESIGN AND SPECIFICATION LANGUAGES FOR EMBEDDED SYSTEMS

Advances in Design and Specification Languages for Embedded Systems

Selected Contributions from FDL'06

Edited by

SORIN A. HUSS
T.U. Darmstadt,
Germany

Springer

A C.I.P. Catalogue record for this book is available from the Library of Congress.

ISBN 978-1-4020-6147-9 (HB)
ISBN 978-1-4020-6149-3 (e-book)

Published by Springer,
P.O. Box 17, 3300 AA Dordrecht, The Netherlands.

www.springer.com

Printed on acid-free paper

Contents

Preface

This book is the latest contribution to the Chip Design Languages series and it consists from selected papers presented at the Forum on Specifications and Design Languages (FDL'06), which took place in September 2006 at Technische Universität Darmstadt, Germany.

FDL, an ECSI conference, is the premier European forum to present research results, to exchange experiences, and to learn about new trends in the application of specification and design languages as well as of associated design and modelling methods and tools for integrated circuits, embedded systems, and heterogeneous systems. Modelling and specification concepts push the development of new methodologies for design and verification to system level, they thus provide the means for a model-driven design of complex information processing systems in a variety of application domains. The aim of FDL is to cover several related thematic areas and to give an opportunity to gain up-to-date knowledge in this fast evolving area. FDL'06 is the ninth of a series of successful events that were held previously in Lausanne, Lyon, Tübingen, Marseille, Frankfurt am Main, and Lille.

Embedded systems are meanwhile in the focus of industry in quite different application domains such as automotive, avionics, telecom, and consumer products. The need for a shift in design methodologies towards system level design is widely recognised and design flows aimed to an integration of software and hardware specification and implementation approaches are being developed. Standardization efforts, such as SystemC Transaction Level Modelling and Model Driven Architecture of the OMG, provide the foundations of these new design flows. Design and specification languages are of utmost interest in the area of embedded systems and the Forum on Specification and design Languages has been once again been the main European event for the embedded systems and chip design community.

This book presents a collection of the best papers from FDL'06, which were selected by the topic area programme chairs Dominique Borrione, Christoph Grimm, Frank Oppenheimer, and Piet van der Putten. The book is structured into four main parts:

Part I – Analog, Mixed-Signal, and Heterogeneous System Design: Design methodologies that exploit a mix of continuous-time and discrete-event modelling languages such as VHDL-AMS, Verilog-AMS, SystemsC–AMS, or Modelica for the design and verification of heterogeneous systems.

Part II – C/C++ Based System Design: Design methodologies that use C/C++ or dedicated modelling languages such as SystemC, SystemVerilog, Verilog, and VHDL jointly with verification languages such as 'e' or PSL/Sugar for the design and verification of hardware/software systems.

Part III – Formalisms for Property-Driven Design: Verification of functional behaviour, generation of test stimuli, model checking on the reachable state space, and direct synthesis.

Part IV – UML-Based System Specification and Design: Specification and design methodologies such as the Model Driven Architecture that rely on UML to map abstract models of complex embedded systems to programmable hardware platforms and to System-on-a-Chip architectures.

The 20 chapters of this book present recent and significant research results in the areas of design and specification languages for embedded systems, SoC, and integrated circuits. I am sure that this book will be a valuable help and reference to researchers, practitioners, and even to students in the field of design languages for electronic components and embedded systems.

Finally, I would like to express my special thanks to Felix Madlener, who put a lot of work into the preparation of this book.

Sorin Alexander Huss
General Chair of FDL'06
Technische Universität Darmstadt
Darmstadt, Germany, December 2006

I

ANALOG, MIXED-SIGNAL, AND HETEROGENEOUS SYSTEM DESIGN

Introduction

The following part of the book focuses the design of analogue and mixed-signal circuits and systems. Compared with the design of digital systems, tools for synthesis are not yet mature or even used in industry. Design of analogue systems is done mostly interactive and done using modelling and simulation. One might think that simulation of analogue circuits with now nearly 50 years of practice and research is mature and stable. However, this is not the case and we see even new challenges. In the following we give a selection of five excellent contributions to hot topics in the modelling, simulation, reuse and verification of analogue and mixed-signal systems.

New technologies like nanotubes are currently emerging and require of course new device models. The first contribution describes the modelling of devices in such emerging technologies and gives the reader an interesting insight into new challenges for at least the next 10 years.

Compared with the design of analogue circuits in the past, we have to analyse the overall behaviour of systems where DSP methods and analogue hardware are interwoven. Today, this is a problem especially when we combine RF components with digital and DSP hardware. The area of baseband modelling is tackled in the second contribution.

Furthermore, system simulation requires behavioural models to get sufficient simulation performance. However, the speed-up of using behavioural models is not yet sufficient. The third contribution describes possible simulator improvements to increase simulation performance of behavioural models.

For the design of analogue systems, there are not yet established tools for synthesis, and analogue design is therefore expensive. An important mean to increase productivity is reuse at different levels of abstraction. The fourth section describes a platform to support the reuse of analogue (and mixed-signal) components by a well-designed web interface with database.

Finally, verification is an important issue – especially with increasing complexity. The fifth contribution describes methods for behavioural modelling with special focus on system verification.

Christoph Grimm

Chapter 1

COMPACT MODELING OF EMERGING TECHNOLOGIES WITH VHDL-AMS

Fabien Prégaldiny[1], Christophe Lallement[1], Birahim Diagne[1], Jean-Michel Sallese[2], and François Krummenacher[2]

[1]*InESS (Institut d'Électronique du Solide et des Systèmes)*
Parc d'innovation, BP 10413
67412 Illkirch Cedex, France
fabien.pregaldiny@iness.c-strasbourg.fr
christophe.lallement@ensps.u-strasbg.fr
birahim@iness.c-strasbourg.fr

[2]*IMM-EPFL*
CH-1015, Lausanne, Switzerland
jean-michel.sallese@epfl.ch
francois.krummenacher@epfl.ch

Abstract This paper deals with the compact modeling of several emerging technologies: first, the double-gate MOSFET (DG MOSFET), and second, the carbon nanotube field-effect transistor (CNTFET). For CNTFETs, we propose two compact models, the first one with a classical behavior (like MOSFET), and the second one with an ambipolar behavior (Schottky-barrier CNTFET). All the models have been compared with numerical simulations and then implemented in VHDL-AMS.

Keywords Compact model, double-gate MOSFET, CNTFET, VHDL-AMS

1. Introduction

Since the introduction of transistors, continuous reduction of electronic circuit size and power dissipation have been the ongoing theme in electronics industry. The well-known "Moore's law" represents this evolution. However,

S.A. Huss (ed.), Advances in Design and Specification Languages for Embedded Systems – Selected Contributions from FDL'06, 5–21.

as the feature size becomes smaller, scaling the silicon MOSFET becomes increasingly harder. This increasing challenge is often attributed to: (1) quantum mechanical tunneling of carriers through the thin gate oxide; (2) quantum mechanical tunneling of carriers from source to drain and from drain to body; (3) control of the density and location of dopant atoms in the channel and source/drain region to provide high on/off current ratio.

There are many solutions proposed to circumvent these limitations. Some solutions include modifications on the existing structures and technologies in hopes of extending their scalability. The DG MOSFET is recognized as one of the most promising candidates for future very large-scale integrated (VLSI) circuits [1, 2]. In DG MOSFETs, short-channel immunity can be achieved with an ideal subthreshold swing (60 mV/dec). Other solutions involve using new materials and technologies to replace the existing silicon MOSFETs. Among them, new device structures as carbon nanotube-based transistors (CNTFETs) are regarded as an important contending device to replace silicon transistors [3].

These new technologies and devices require the creation of accurate compact models, suited to the circuit design and easily translatable into a hardware description language (HDL) such as VHDL-AMS.

This paper is organized as follows. In Section 2, we present an explicit model for the symmetric DG MOSFET that is simple, inherently continuous, and computationally efficient. By introducing useful normalizations as in the EKV MOSFET model, we have derived simple and clear relationships which are really helpful for the circuit designer [4]. In Section 3, we propose two compact models for CNTFETs, the first one with a conventional behavior (i.e. a MOSFET behavior), and the second one with an ambipolar behavior. The former is based on an existing model developed at Purdue University [5]. Unfortunately, in its present form, this model is not appropriate for circuit simulation. In this paper, we propose an efficient compact model for the designer, with a range of validity clearly defined. The second model is devoted to compact modeling of the CNTFET with an ambipolar behavior (n- or p-type depending of the gate voltage value). This characteristic is quite different from a classic behavior, namely a MOSFET behavior. To our best knowledge, this compact model is the first analytical ambipolar model for CNTFET introduced in the literature. It is a behavioral compact model that simulates in a realistic way the ambipolar characteristic observed with Schottky-Barrier (SB) CNTFETs.

2. Double-Gate MOSFET

A Compact Model Dedicated to the Design

For the last decade, a significant research effort in this field has led to the development of physical models for the DG MOSFET [6–8]. These models are of major interest for the design of the device itself but less useful for circuit simulation since they rely on very complicated formulations. Among the proposed models, Taur's model [9] is one of the best candidates for building a compact model. An exact solution for both charges and current has been proposed and successfully validated. However, such a model, in its current form, is not really suited for circuit simulation because it requires an iterative procedure to compute the mobile charge density, which is generally considered to be time consuming.

The main assumptions of our new model are the following: the body (i.e. the silicon layer) is undoped or lightly doped, the mobility is constant along the channel and both quantum effects and polydepletion effect are neglected. The last assumption is valid for silicon layer thicknesses down to at least 20 nm. For thinner layers, quantum effects start to play a role [6, 8], but might actually be considered as a correction to the classical derivation. The schematic diagram of the DG MOSFET considered in this work is shown in Fig. 1.1.

Using the normalization of charges, potentials, and current proposed in [4] leads to an important relationship between charge densities and potentials, given by

$$v_g^* - v_{ch} - v_{to} = 4 \cdot q_g + \ln q_g + \ln\left(1 + q_g \cdot \frac{C_{ox1}}{C_{si}}\right) \tag{1.1}$$

where v_g^* is the effective gate voltage ($= v_g - \Delta\phi_i$ with $\Delta\phi_i$ the work function difference between the gate electrode and intrinsic silicon), v_{ch} is the electron quasi-Fermi potential, v_{to} is the threshold voltage, q_g is the charge density per unit surface of each gate, C_{ox1} is the gate oxide capacitance per unit surface of each gate and C_{si} is the silicon layer capacitance per unit surface.

Such a normalization represents an efficient tool for the analog designer because it is done taking into account the design methodologies requirements [10]. However, (1.1) needs to be solved numerically and this is not desirable

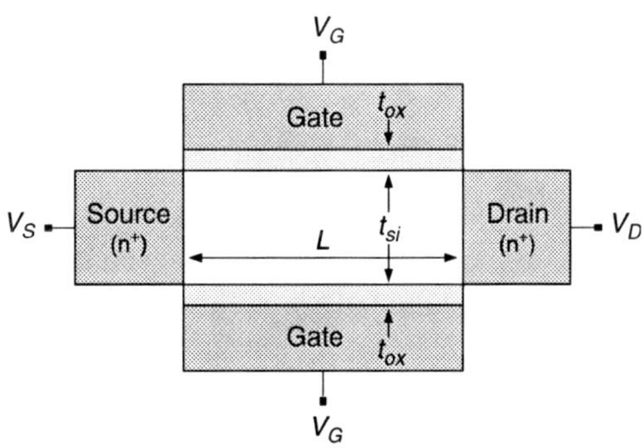

Figure 1.1. Schematic of the DG MOSFET structure.

for circuit simulation (it requires at least several iterations). To overcome this drawback, we have developed a new methodology to compute without any iteration the mobile charge density as an explicit function of bias voltages (v_g and v_d or v_s) [11]

$$q_g = f(v_g, v_{ch}) \text{ with } v_{ch} = v_s \text{ or } v_d \tag{1.2}$$

Without entering into details, the numerical inversion of (1.1) can be performed using a reduced set of precomputed parameters that depend only on the "form factor" α ($= C_{ox1}/C_{si}$). Let us emphasize that our algorithm of numerical inversion fully preserves the physics of (1.1), and therefore its validity is technology-independent [11].

Then, noting that the mobile charge density is twice the gate charge density ($q_m = -2\,q_g$) and assuming that the drift-diffusion transport model is valid, the normalized drain current i can be expressed as

$$i = -\int_{v_s}^{v_d} q_m \cdot dv_{ch} \tag{1.3}$$

Integrating (1.3) from source to drain yields

$$i = -q_m^2 + 2 \cdot q_m + \frac{2}{\alpha} \cdot \ln\left(1 - \frac{\alpha \cdot q_m}{2}\right)\Bigg|_{q_{ms}}^{q_{md}} \tag{1.4}$$

Finally, the drain current I_D is obtained after denormalization of (1.4) as outlined in [11].

To conclude this brief description of the model, it should be said that in addition to the static part, a more complete compact model should include the dynamic part, i.e. the transconductances and the whole set of transcapacitances. The derivation of the dynamic model is not within the scope of this paper and the reader is referred to references [12, 13] for full details. However, let us emphasize that the VHDL-AMS code of our model includes both static and dynamic models [11].

VHDL-AMS Implementation

VHDL-AMS [14] is an HDL which supports the modeling and the simulation of analog and mixed-signal systems. It supports the description of continuous-time behavior. For compact modeling, the most interesting feature of the language is that it provides a notation for describing Differential Algebraic Equations (DAEs) in a fairly general way [15]. The == operator and the way the quantities (bound to terminals or free) are declared allow the designer to write equations in either implicit or explicit format. VHDL-AMS supports the description of networks as conservative-law networks (Kirchhoff's networks) and signal-flow networks (inputs with infinite impedance, outputs with zero impedance). As such, it supports the description and the simulation of multi-discipline systems at these two levels of abstraction. Conservative-law relationships assume the existence of two classes of specialized quantities, namely *across quantities* that represent an effort (e.g. a voltage for electrical systems), and *through quantities* that represent a flow (e.g. a current for electrical systems).

Listing 1 presents the *entity* part of the VHDL-AMS code for the DG MOSFET model. The code first contains references to libraries needed to parse the model (lines 1–3). For the model end user (circuit designer), the most important part of the model is the interface, contained in what is called an `entity` in VHDL-AMS (lines 4–11).

```
(1)  library ieee; library disciplines;
(2)  use disciplines.electromagnetic_system.all;
(3)  use work.all;
(4)  entity dg_mosfet is
(5)    generic(W    :real:= 1.0e-6; -- Gate width [m]
(6)            L    :real:= 1.0e-6; -- Gate length [m]
(7)            tox1 :real:= 2.0e-9; -- Gate oxide thickness [m]
(8)            tsi  :real:= 25.0e-9; -- Si film thickness [m]
(9)            mu0  :real:= 0.1); -- Low-field mobility [m^2/Vs]
(10)   port (terminal g1,g2,d,s :electrical);
(11) end;
```

Listing 1. Interface of the DG MOSFET VHDL-AMS model: the *entity*.

The model interface includes the specification of generic parameters (lines 5–9) and interface ports (line 10). The `generic` statement allows the designer to define its own values for the model parameters. Typically, geometrical W and L transistor parameters are defined as generic. The `dg_mosfet` entity contains four terminals (g1, g2, d, and s stand for the top gate, bottom gate, drain, and source terminal, respectively), all of electrical type. All the terminals are part of a `port` statement. The second part of the VHDL-AMS code is self-explicit. The device behavior is defined in an architecture named symmetric (130 lines of code [11]).

Results and Discussion

To conclude this section, we present the results obtained with the VHDL-AMS simulations of the DG MOSFET model. Figure 1.2 illustrates the computation of the drain current I_D at $V_{DS} = 50$ mV and 1 V. The VHDL-AMS simulation gives evidence for the good numerical behavior of the model in all regions of operation. In particular, the phenomena of volume inversion (i.e. the weak-inversion region) is well described.

Figure 1.3 shows a common set of normalized transcapacitances (with respect to $C_{OX} = 2WLC_{ox1}$) versus the gate voltage. An important point is that all

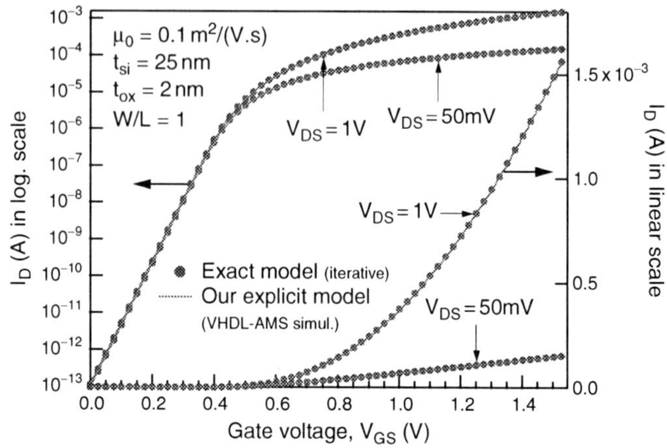

Figure 1.2. Comparison between the results extracted from a VHDL-AMS simulation and the exact iterative model (dots, cf. [9]) for the drain current I_D vs. V_{GS} of a symmetrical DG MOSFET.

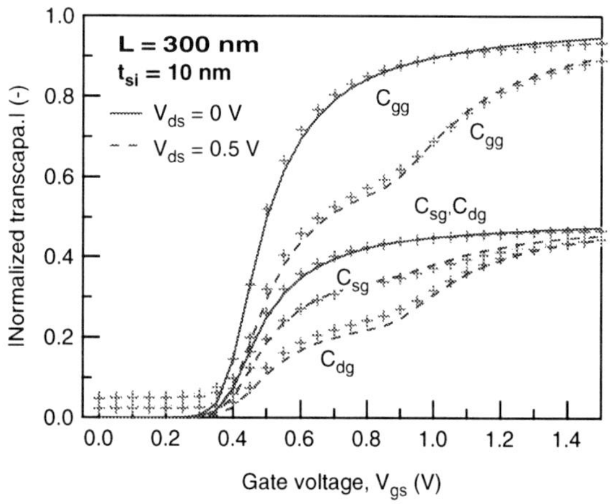

Figure 1.3. Comparison between the *C–V* curves obtained by the explicit model (lines) and 2D simulations (markers).

transcapacitances are continuous between all operating regions without using any fitting parameter, which makes our explicit model numerically robust as well as close to physics. It appears that the model predictions are accurate and fit the 2D simulations in all cases, namely at $V_{DS} = 0$ and $V_{DS} \neq 0$. The slight deviation in the subthreshold region results from the increasing influence of the overlap capacitance as the channel length decreases. For devices with $L \geqslant 1\ \mu$m, the overlap capacitance is negligible. A further development of the model will include the extrinsic capacitances.

3. CNTFETs

Carbon nanotubes (CNTs) are currently considered as promising building blocks of a future nanoelectronic technology. CNTs are hollow cylinders composed of one or more concentric layers of carbon atoms in a honeycomb lattice arrangement. Single-walled nanotubes (SWCNTs) typically have a diameter of 1–2 nm and a length up to several micrometers. The large aspect ratio makes the nanotubes nearly ideal one-dimensional (1D) objects, and as such the SWCNTs are expected to have all the unique properties predicted for these low-dimensional structures [3]. In addition, depending on the detailed arrangement of the carbon atoms the SWCNTs can be metallic or semiconducting. Two types of semiconducting CNTs are being extensively studied. One of these devices is a tunneling device, shown in Fig. 1.4(a). It works on the principle of direct tunneling through a Schottky barrier at the source-channel (and drain-channel) junction. The barrier width is modulated by the application of gate voltage and thus the transconductance of the device is dependent on the gate voltage. To overcome these handicaps associated with the SB CNTFETs, there have been attempts to develop CNTFETs which would behave like normal MOSFETs [Fig. 1.4(b)]. In this MOSFET-like device, the ungated portion (source and drain regions) is heavily doped and the CNTFET operates on the principle of barrier-height modulation by application of the gate potential. In this case, the on-current is limited by the amount of charge that can be induced in the channel by the gate. It is obvious that the MOSFET-like device will give a higher on-current and, hence, would define the upper limit of performance.

Transport through short nanotubes has been shown to be free of significant acoustic and optical phonon scattering and thus is essentially ballistic at both high and low voltage limits. In the following, we consider MOSFET-like mode of operation, and assume ballistic transport.

The theory of CNT transistors is still primitive and the technology is still nascent. However, evaluation of such high-performance transistors in digital

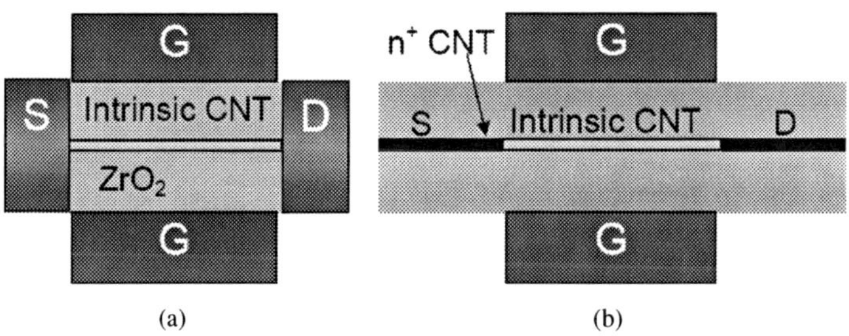

Figure 1.4. Different types of CNTFETs: (a) Schottky-barrier (SB) CNTFET with ambipolar behavior, and (b) MOSFET-like CNTFET with classic behavior.

circuits is absolutely essential to drive the device design and devise new architectures. However, from the circuit designer's point of view, circuit simulation and evaluation using CNTFETs is challenging because most of the developed models are numerical, involving self-consistent equations which circuit solvers like SPICE are not able to handle.

MOSFET-like CNTFET

First, we present a compact model for CNTFETs with a classical behavior. This compact model is based on a CNTFET model developed at Purdue University [5]. To our best knowledge, Purdue's model is the first *compact model* (i.e. fully dedicated to circuit simulation) of CNTFET available in the literature. It is a surface potential-based SPICE compatible model that enables to simulate CNTs with ballistic behavior. It has been incorporated in HSPICE but is not well-suited for circuit simulation due to some convergence issues.

In this paper, we propose a modified model with fundamental improvements solving the convergence problems of the original model. The new model is applicable to a wide range of CNTFETs with diameters between 1 and 3 nm and for all chiralities as long as they are semiconducting. The model uses suitable approximations necessary for developing any quasi-analytical, circuit-compatible compact model (see Fig. 1.5). Quasi-static characteristics (*I–V*) have been modeled and validated against numerical models, with an excellent agreement.

The computional procedure to evaluate the drain current I_D and the total channel charge Q_{CNT} is illustrated in Fig. 1.6. The main quantities used in the model are the surface potential ψ_S (or control potential) and the specific voltage $\xi_{S(D)}$ that depends on the surface potential, the subbands energy level Δ_p and the source (drain) Fermi level $\mu_{S(D)}$. The conduction band minima for the first subband is set to half the nanotube bandgap Δ_1 with $\Delta_1 \simeq 0.45/\text{diam}$ (in eV).

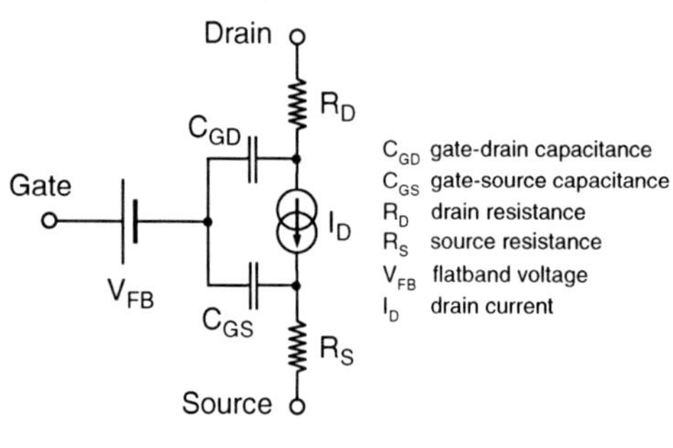

Figure 1.5. Schematic of the CNTFET compact model.

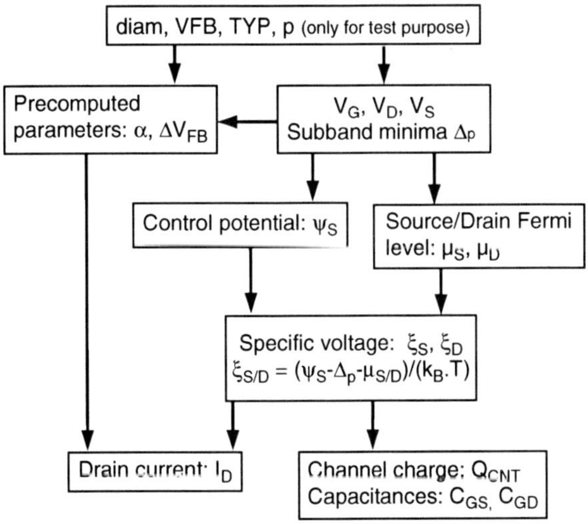

Figure 1.6. Structure of the CNTFET compact model.

The physical parameter diam is the nanotube diameter (in nm); it is one of the only three intrinsic parameters of our model, with the flatband voltage V_{FB} and the TYP parameter ($= +1/-1$ for n- or p-type device). Let us emphasize the number of subbands p has been added as an input parameter only for test purpose [16].

Determination of the surface potential. An important step in the model development is to relate the control potential with the gate bias voltage (see Fig. 1.6). The knowledge of ψ_S is useful to calculate the specific voltage ξ. This allows us to determine the drain current and the total charge. In [5], the following approximation has been proposed

$$V_{GS} - \psi_S = \begin{cases} 0 & \text{for } V_{GS} < \Delta_1, \\ \alpha \cdot (V_{GS} - \Delta_1) & \text{for } V_{GS} \geqslant \Delta_1. \end{cases} \tag{1.5}$$

where the parameter α is given by

$$\alpha = \alpha_0 + \alpha_1 \cdot V_{DS} + \alpha_2 \cdot V_{DS}^2 \tag{1.6}$$

where α_0, α_1, and α_2 are dependent on both CNTFET diameter and gate oxide thickness [16]. Eq. (1.5) is correct to model the relationship between the gate voltage and the surface potential, but is not well-suited for a compact model (problem of discontinuity, as shown in Fig. 1.7). Therefore, we propose

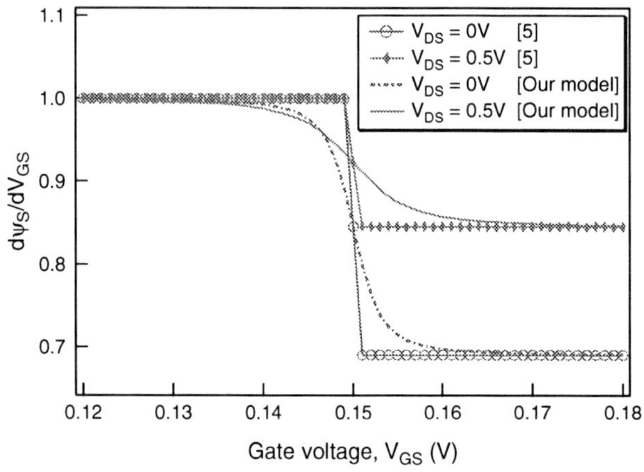

Figure 1.7. Derivative of surface potential ψ_S vs. V_{GS}.

an equivalent solution, given by (1.7), but with an excellent behavior of the derivative (see Fig. 1.7)

$$\psi_S = V_{GS} - \frac{\alpha(V_{GS} - \Delta_1) + \sqrt{[\alpha(V_{GS} - \Delta_1)]^2 + 4\epsilon^2}}{2} \tag{1.7}$$

where $\epsilon = 5 \cdot 10^{-2}$ is a smoothing parameter.

Then, the total drain current I_D is obtained as

$$I_D = \frac{4\,q\,k_B\,T}{h} \sum_p \left[\ln\left(1 + \exp(\xi_S)\right) - \ln\left(1 + \exp(\xi_D)\right)\right] \tag{1.8}$$

where p is the number of subbands, k_B and h are the constants of Boltzmann and Planck, respectively.

Quantum-Capacitance Derivation. With the knowledge of charge and surface potential as functions of gate bias, the gate input capacitance C_G can be computed in terms of the device parameters and terminal voltages. The gate input capacitance is given by

$$C_G = \frac{\partial Q_{CNT}}{\partial V_{GS}} \quad \Rightarrow \quad C_G = \frac{\partial Q_{CNT}}{\partial \psi_S} \cdot \frac{\partial \psi_S}{\partial V_{GS}} \tag{1.9}$$

The total charge Q_{CNT} can be split up into Q_S and Q_D and, hence, the total gate capacitance can also be split up into C_{GS} and C_{GD} (see Fig. 1.5).

To elaborate an efficient expression of C_G for a compact model, it is important to first have a closed-form expression of $Q_{CNT}(\psi_S)$ and continuous derivatives of (1.9) as well. As it is not possible to obtain a closed-form relationship for the quantum-charge in the channel, an empirical solution (fit) has been proposed in [5]. Noting that the number of carrier n increases almost linearly as ξ increases and falls off exponentially as ξ becomes negative, the following relationship has been derived

$$n = \begin{cases} N_0 \cdot A \cdot \exp \xi & \text{for } \xi < 0, \\ N_0 \cdot (B \cdot \xi + A) & \text{for } \xi \geqslant 0. \end{cases} \tag{1.10}$$

where the parameters A and B are dependent on the energy level Δ [5].

Eq. (1.10) is unfortunately not appropriate for circuit simulation because its derivatives are not continuous (Fig. 1.8). Accordingly, the different capacitances determined by (1.10) would not be correct to elaborate the CNTFET dynamic model. In addition, this would lead to numerical problems during simulation and wrong results. In order to solve the numerical problems, we have elaborated a new equation for n, similar to the interpolation function of the EKV MOSFET model [10]. This new expression and its derivatives (Fig. 1.8) are continuous and well-suited for circuit simulation, especially in dynamic operation

$$n_{\text{new}}(\xi) = N_0 \cdot 1.2 \cdot B \cdot \left\{ \ln \left[1 + \frac{A}{1.2 \cdot B} \cdot \exp\left(\frac{\xi}{0.96}\right) \right] \right\}^{0.96} \tag{1.11}$$

Figure 1.8 shows a comparison between the derivatives of (1.10) and (1.11). Let us note that the greatest difference can be observed around zero, where actually the former overestimates the quantum-charge (see Fig. 4 in [5]). The VHDL-AMS simulation of the capacitances computed with our continuous model is shown in Fig. 1.9.

Figure 1.10 shows the drain current of a 1.4 nm diameter CNTFET with $C_{ox} = 3.8$ pF/cm as a function of gate voltage. The dots correspond to the

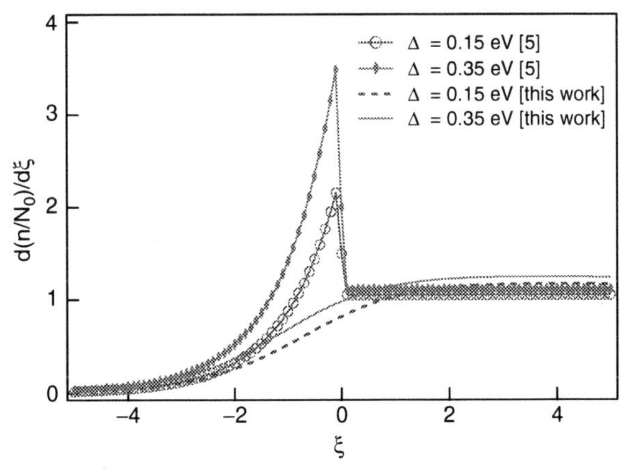

Figure 1.8. Improvement of the numerical behavior.

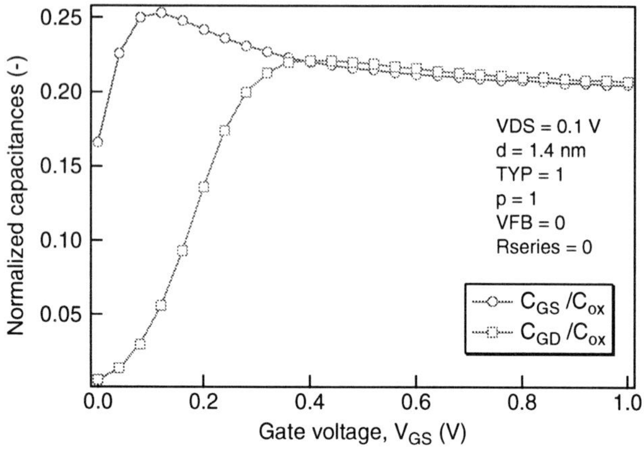

Figure 1.9. VHDL-AMS simulation of the C_{GS} and C_{GD} capacitances for a MOSFET-like CNTFET with $C_{ox} = 3.8$ pF/cm.

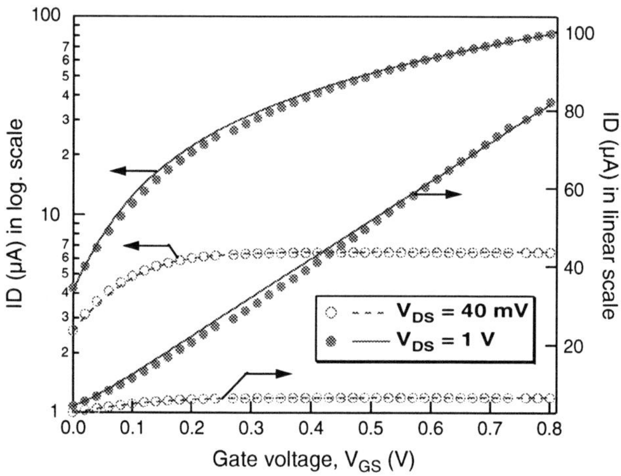

Figure 1.10. Comparison between the results extracted from VHDL-AMS and numerical simulations (lines and dots, respectively) for the drain current of a MOSFET-like CNTFET (p = 1, diam = 1.4, TYP = +1, VFB = 0, Rseries = 0).

numerical solutions performed with the FETToy simulator [17] whereas the lines correspond to our analytical compact model. A good agreement is found, which supports the validity of our approach.

VHDL-AMS Implementation

First, we have calibrated the model of Purdue with respect to numerical simulations [18, 17]. The best fits were obtained with $p = 1$ (i.e. one subband) which is coherent because the FETToy simulator only accounts for the lowest subband. So, at the beginning, we fixed $p = 1$ in our model in order to validate it with respect to the numerical simulations. Then, if we consider CNTFETs with diameters ranging from 1 to 3 nm, and with a power supply lower than 1 V, we can set $p = 5$ to accurately describe all cases [16].

The whole VHDL-AMS code of the model requires about 90 lines. Only three intrinsic parameters are necessary: diam, TYP ($+1$ for n-type, -1 for p-type) and V_{FB} (lines 5–7 in Listing 2).

```
(1)  library ieee; library disciplines;
(2)  use disciplines.electromagnetic_system.all;
(3)  use work.all;
(4)  entity CNTFET is
(5)    generic(diam    : real := 1.4; -- Nanotube diameter [nm]
(6)            TYP     : real := 1.0; -- n/p-CNTFET (+1/-1)
(7)            VFB     : real := 0.0; -- Flatband voltage [V]
(8)            p       : positive := 1; -- Number of subbands [-]
(9)            Rseries : real := 50.0e3); -- Series resistance [ohm]
(10)   port(terminal g,d,s : electrical);
(11) end;
```

Listing 2. Interface of the CNTFET VHDL-AMS model: the *entity*.

Let us note that the number of subbands p has been defined as a generic parameter only for test purpose [16]. Rseries corresponds to the total series resitance, that is $R_{\text{source}} + R_{\text{drain}}$ with $R_{\text{source}} = R_{\text{drain}}$. The parameters α_0, α_1, and α_2 [see (1.6)] are determined in a precomputed module, with the help of one equation for each of them. For all details about the computation of the parameters α, the reader is referred to [16].

To conclude this section, Fig. 1.11 shows two VHDL-AMS simulations performed for different values of the parameters diam and p, in order to show the effect of the nanotube diameter on the number of subbands p to be accounted for. This behavior may be useful to create novel multiple-valued logic design [19].

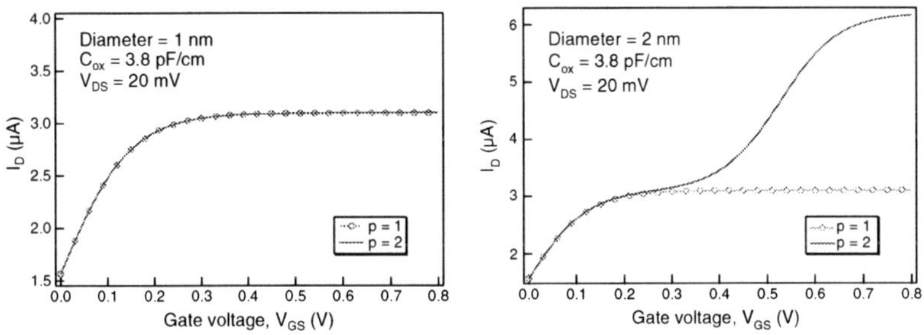

Figure 1.11. VHDL-AMS simulations of I_D vs. V_{GS} at low drain bias.

Ambipolar CNTFET

We present, for the first time to our best knowledge, a behavioral compact model that allows to describe the ambipolar characteristic of SB CNTFETs. This model is built using the new model of CNTFET previously presented. As shown in Fig. 1.12, an additional part has been added to the "unipolar" model. The entity (VHDL-AMS) corresponding to this new model is the same as the classical CNTFET model one.

The very particular I_D–V_{GS} characteristic of the ambipolar CNTFET is illustrated in Fig. 1.13. It should be noted that this behavior is quite similar to the numerical simulation results recently published in [20] and [21]. This

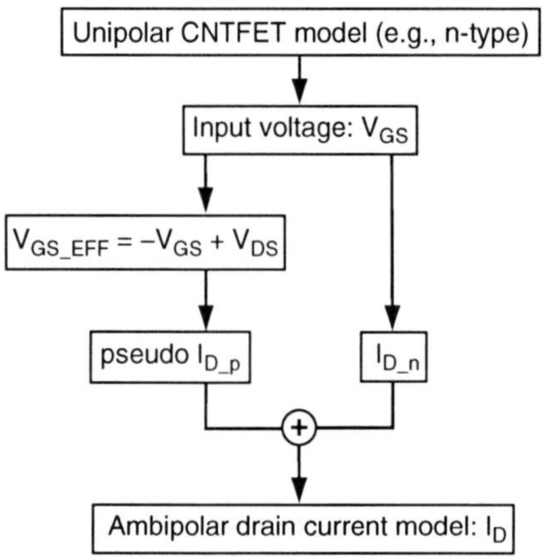

Figure 1.12. Structure of the behavioral model for the ambipolar CNTFET.

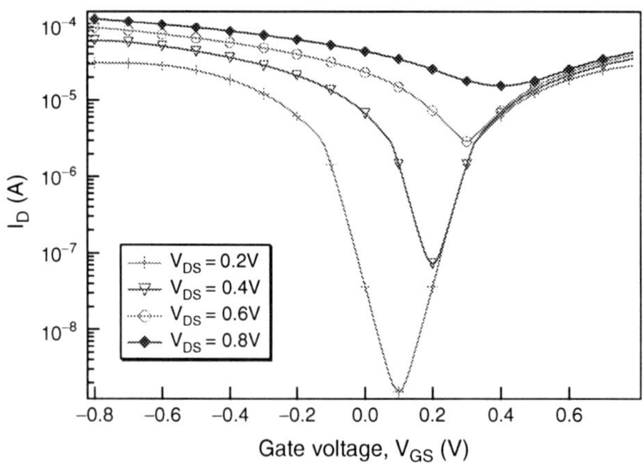

Figure 1.13. VHDL-AMS simulation of the drain current as a function of gate voltage for the ambipolar SB CNTFET (p = 1, diam = 1.4, TYP = +1, VFB = 0, Rseries = 0).

ambipolar characteristic should allow circuit designers to devise new architectures using that specific behavior [22, 23]. Our compact model may be of help to this issue.

4. Conclusion

In this paper, different VHDL-AMS models for emerging technologies have been proposed. The DG MOSFET model is still under development, it will be completed with the modeling of additional effects such as quantum effects, extrinsic capacitances, in order to simulate accurately ultra short-channels DG MOSFETs. The second part of the paper dealt with the compact modeling of the CNTFET with VHDL-AMS. Two CNTFET compact models have been presented, the first one for carbon nanotubes with a classical behavior (like MOSFET), and the second one for devices with an ambipolar behavior. Although CNTFET technology is still nascent, these compact models developed in VHDL-AMS are useful tools to help the designers to devise new architectures.

Acknowledgment

This work was partially funded by the French Ministry of Research under Nanosys ("Programme ACI Nanosciences").

References

[1] D.J. Frank, R.H. Dennard, E. Nowak, P.M. Solomon, Y. Taur, and H.-S.P. Wong. Device scaling limits of Si MOSFETs and their application dependencies. *Proc. of IEEE*, 89(3):259–288, 2001.

[2] International technology roadmap for semiconductors 2004 update. [Online]. Available: http://public.itrs.net/ [May 2005].

[3] P. Avouris, J. Appenzeller, R. Martel, and S.J. Wind. Carbon nanotube electronics. *Proc. of IEEE*, 91(11):1772–1784, 2003.

[4] J-M. Sallese, F. Krummenacher, F. Prégaldiny, C. Lallement, A. Roy, and C. Enz. A design oriented charge-based current model for symmetric DG MOSFET and its correlation with the EKV formalism. *Solid-State Electron.*, 49(3):485–489, 2005.

[5] A. Raychowdhury, S. Mukhopadhyay, and K. Roy. A circuit-compatible model of ballistic carbon nanotube field-effect transistors. *IEEE Trans. Computer-Aided Design*, 12(10):1411–1420, 2004.

[6] L. Ge and J.G. Fossum. Analytical modeling of quantization and volume invesion in thin Si-film double-gate MOSFETs. *IEEE Trans. Electron Devices*, 49(2):287–294, 2002.

[7] Y. Taur. An analytical solution to a double-gate MOSFET with undoped body. *IEEE Electron Device Lett.*, 21(5):245–247, 2000.

[8] J.G. Fossum, L. Ge, M-H. Chiang, V.P. Trivedi, M.M. Chowdhury, L. Mathew, G.O. Workman, and B-Y. Nguyen. A process/physics-based compact model for nonclassical CMOS device and circuit design. *Solid-State Electron.*, 48:919–926, 2004.

[9] Y. Taur, X. Liang, W. Wang, and H. Lu. A continuous, analytic drain-current model for DG MOSFETs. *IEEE Electron Device Lett.*, 25(2): 107–109, 2004.

[10] C. Enz, F. Krummenacher, and E. Vittoz. An analytical MOS transistor model valid in all regions of operation and dedicated to low-voltage and low-current applications. *Journal of AICSP*, pp. 83–114, 1995.

[11] F. Prégaldiny, F. Krummenacher, B. Diagne, F. Pêcheux, J-M. Sallese, and C. Lallement. Explicit modelling of the double-gate MOSFET with VHDL-AMS. *Int. Journ. of Numerical Modelling: Elec. Networks, Devices and Fields*, 19(3):239–256, 2006.

[12] J-M. Sallese and A.S. Porret. A novel approach to charge-based non-quasi-static model of the MOS transistor valid in all modes of operation. *Solid-State Electron.*, 44:887–894, 2000.

[13] F. Prégaldiny, F. Krummenacher, J-M. Sallese, B. Diagne, and C. Lallement. An explicit quasi-static charge-based compact model for symmetric DG MOSFET. In: *NSTI-Nanotech 2006, WCM*, vol. 3, pp. 686–691, 2006. ISBN 0-9767985-8-1.

[14] *1076.1-1999 IEEE Standard VHDL Analog and Mixed-Signal Extensions Language Reference Manual*, IEEE Press edition, 1999. ISBN 0-7381-1640-8.

[15] F. Pêcheux, C. Lallement, and A. Vachoux. VHDL-AMS and Verilog-AMS as alternative hardware description languages for efficient modelling of multi-discipline systems. *IEEE Trans. Computer-Aided Design*, 24(2):204–224, 2005.

[16] F. Prégaldiny, C. Lallement, and J.-B. Kammerer. Design-oriented compact models for CNTFETs. In: *Proc. of IEEE DTIS'06*, pp. 34–39, 2006.

[17] NANOHUB online simulations and more, CNT bands. [Online]. Available: http://www.nanohub.org [April 2006].

[18] J. Guo, M. Lundstrom, and S. Datta. Performance projections for ballistic carbon nanotube field-effect transistors. *App. Phys. Letters*, 80(17): 3192–3194, 2002.

[19] A. Raychowdhury and K. Roy. Carbon-nanotube-based voltage-mode multiple-valued logic design. *IEEE Trans. Nanotechno.*, 4(2):168–179, 2005.

[20] J. Knoch, S. Mantl, and J. Appenzeller. Comparison of transport propoerties in carbon nanotube field-effect transistors with Schottky contacts and doped source/drain contacts. *Solid-State Electron.*, 49:73–76, 2005.

[21] J. Guo, S. Datta, and M. Lundstrom. A numerical study of scaling issues for schottky-barrier carbon nanotube transistors. *IEEE Trans. Electron Devices*, 51(2):172–177, 2004.

[22] R. Sordan, K. Balasubramanian, M. Burghard, and K. Kern. Exclusive-OR gate with a single carbon nanotube. *App. Phys. Letters*, 88, 2006. 053119.

[23] F. Prégaldiny, C. Lallement, and J.-B. Kammerer. Compact modeling and applications of CNTFETs for analog and digital circuit design. In: *Proc. of IEEE ICECS'06,* pp. 1030–1033, 2006.

Chapter 2

BASEBAND MODELING USING MULTIDIMENSIONAL NETWORKS IN VHDL-AMS

Joachim Haase
Fraunhofer-Institut Integrierte Schaltungen/Branch Lab Design Automation EAS
Zeunerstr. 38, D-01069 Dresden, Germany
Joachim.Haase@eas.iis.fraunhofer.de

Abstract Baseband models are widely used to describe the behavior of RF systems. They suppress the RF carrier. Thus, they are many times faster than passband models that evaluate every carrier cycle during the simulation. Nevertheless, passband models are necessary for the component design. To support the top-down design approach, consistent models at baseband and passband level are required. The paper shows how the consistency can be achieved using consequently the possibilities of overloading functions and operators in VHDL-AMS. The mathematical basis is a consistent usage of the describing function theory for baseband modeling. The classical approach of baseband modeling can be extended by completing the waveforms that carry the in-phase and quadrature component by a time-discrete or time-continuous waveform that saves the carrier frequency information. This allows to handle systems with different carriers in different parts of a system or to sweep the carrier frequency during the simulation.

Keywords VHDL-AMS, baseband modeling, describing function

1. Introduction

Standard languages as Verilog-A, Verilog-AMS, and VHDL-AMS [1] are available to model the behavior of mixed-signal circuits and systems. They provide greater understanding of systems early in the design process. In order to be able to compare different system architectures a high execution speed of the simulation is required.

S.A. Huss (ed.), Advances in Design and Specification Languages for Embedded Systems – Selected Contributions from FDL'06, 23–35.

Baseband models fulfill these requirements for a wide class of RF systems [3, 7]. They are not as accurate as passband models but they are much faster. The idea behind baseband descriptions is that narrow band analog waveforms with the carrier frequency f can be represented by the following equation

$$x(t) = A(t) \cdot \cos(\omega t + \varphi(t)) = I(t) \cdot \cos(\omega t) - Q(t) \cdot \sin(\omega t) \quad (2.1)$$

that means

$$x(t) = Re\left((I(t) + j \cdot Q(t))\, e^{j\omega t}\right) \quad (2.2)$$

where $\omega = 2\pi \cdot f$ and Re gives access to the real part of a complex number. All the waveforms x, A, f, I, and Q are scalar real-valued time-continuous waveforms. A special waveform $x(t)$ is characterized by a 3 tupel $(I(t), Q(t), f)$. x is called the passband representation and the 3 tupel carries the information of the baseband representation of the waveform. The set of all waveforms that can be described in this way shall be indicated by $X_{PB}(f)$ for the passband representation and $X_{BB}(f)$ for the baseband representation. The sum $I(t) + j \cdot Q(t)$ be interpreted as a slowly varying phasor [3]. In the following we will try to define the baseband operations so that passband functionality can be mapped in an easy way to the baseband description.

The next section introduces the mapping in a formal way. Afterward, the implementation in VHDL-AMS is described and illustrated with the help of two examples.

2. Baseband Modeling Approach

Linear operations as addition, subtraction, multiplication with a constant, and differentiation (see [10], [3]) in the passband can be easily assigned to linear operations in the baseband.

Functions Without Frequency Shifting

Considering a unary nonlinear passband map m_{PB}, then in the general case the range can be built up by the union of the sets characterized by the fundamental, its harmonics, and a static contribution

$$m_{PB} : X_{PB}(f) \rightarrow X_{PB}(0) \cup X_{PB}(f) \cup X_{PB}(2 \cdot f) \cup \ldots \quad (2.3)$$

Subharmonics are not considered. Baseband modeling usually introduces the simplification that the associated baseband map m_{PB} only describes the mapping onto the set of fundamentals. That means

$$m_{BB} : X_{BB}(f) \rightarrow X_{BB}(f) \quad (2.4)$$

The modeling decision is whether this simplification can be accepted or not. Many linear and nonlinear maps can be represented using describing functions [5]. The describing function is the phasor representation of the output of a scalar (nonlinear) map at frequency f divided by the phasor representation of the argument of the map at frequency f.

Modeling procedure. Assume the describing function $N(A, f) \in \mathbb{C}$ can be assigned to a passband map m_{PB} then the associated baseband map m_{BB} can be carried out in the following way

$$\begin{aligned} & m_{BB} : X_{BB}(f) \rightarrow X_{BB}(f) \\ & \text{with} \\ & (I(t), Q(t), t) \\ & \mapsto (Re(N(A(t), f) \otimes B(t)), Im(N(A(t), f) \otimes B(t)), f) \end{aligned}$$

with $A(t) = \sqrt{I(t)^2 + Q(t)^2}$, $B(t) = I(t) + j \cdot Q(t)$, the multiplication sign $\otimes$ in the complex area and access to the real and imaginary part of a complex number with Re and Im resp. □

Example 1

Figure 2.1 shows a nonlinear characteristic that can be used for modeling of low noise amplifiers [3].

$$\tilde{m}_{PB} : x(t) \mapsto \begin{cases} -A_{max} & \text{for } x(t) \leq A_{lim} \\ A_1 \cdot x(t) - A_3 \cdot x(t)^3 & \text{for } others \\ A_{max} & \text{for } x(t) \geq A_{lim} \end{cases}$$

with the real numbers A_1, A_3, $A_{lim} = \sqrt{\frac{A_1}{3 \cdot A_3}}$ and $A_{max} = \frac{2}{3} \cdot A_1 \cdot A_{lim}$. The describing function based on [5] is given by

$$N(A, f) = \begin{cases} \frac{2}{3 \cdot A} \cdot \left(\tilde{m}_{PB}(A) + \tilde{m}_{PB}\left(\frac{A}{2}\right)\right) & \text{for } A > A_{lim} \\ A_1 - \frac{3}{4} \cdot A_3 \cdot A^2 & \text{for } others. \end{cases}$$

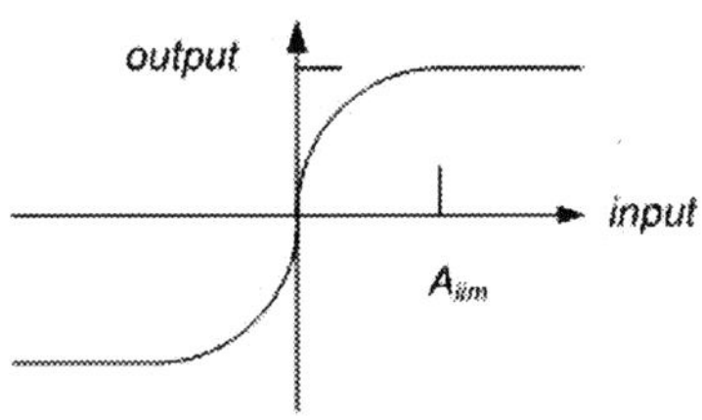

Figure 2.1. Graph of function $\tilde{m}_{PB}$.

Thus, the baseband characteristic can easily be expressed using the procedure introduced above $(N(A, f) \in \mathbb{R})$

$$\begin{aligned} &\tilde{m}_{BB} : X_{BB}(f) \to X_{BB}(f) \text{ with} \\ &(I(t), Q(t), t) \mapsto (N(A, f) \cdot I(t), N(A, f) \cdot Q(t) f) \end{aligned}$$

where $A(t) = \sqrt{I(t)^2 + Q(t)^2}$. This approach seems to be more convenient than only clamping the output amplitude for higher input amplitudes (cf. e.g. [3]) □

It should be considered that using this approach baseband models for several functions as quantizers, general piecewise-linear nonlinearities, limiters, exponential saturation, algebraic saturation, ideal hysteresis functions, and others can be established in a very simple way. This approach also offers the opportunity to start in some of these cases with table-look up descriptions (see also [8]).

Frequency Shifting

Looking at equation (2.3) it should be mentioned that also $X(0)$ and $X(n \cdot f)$ (with $\mathrm{n} = 2, 3,$ Ė) are possible ranges of a baseband map. This way down- or up-conversion of the carrier frequency between the system's parts can be handled.

Example 2
In the case of a multiplier two alternatives exist. The passband map is described by

$$\bar{m}_{PB} : X_{PB}(f) \times X_{PB}(f) \to X_{PB}(0) \cup X_{PB}(2 \cdot f)$$

with

$$\begin{aligned} &(A_1(t) \cdot \cos(\omega t + \varphi_1(t)), A_2(t) \cdot \cos(\omega t + \varphi_2(t))) \\ &\mapsto \tfrac{1}{2} \cdot A_1(t) \cdot A_2(t) \cdot (\cos(\varphi_1(t) - \varphi_2(t)) + \cos(2\omega t + \varphi_1(t) + \varphi_2(t))) \end{aligned}$$

and $\omega = 2\pi f$. Thus, the down-conversion baseband characteristic is

$$\begin{aligned} &\bar{m}_{BB} : X_{BB}(f) \times X_{BB}(f) \to X_{BB}(0) \text{ with} \\ &((I_1, Q_1, f), (I_2, Q_2, f)) \mapsto \left(\tfrac{1}{2}(I_1 \cdot I_2 + Q_1 \cdot Q_2), 0, 0\right) \square \end{aligned}$$

Baseband modeling can be applied in an easy way if the carrier frequencies of the operands of the baseband characteristics are equal. To shift a representation with carrier frequency f_1 to a representation with carrier frequency f_2 the following equation for the passband representation should be considered (compare (2.2))

$$\begin{aligned} &Re\left(\left(I\left(t\right)+j\cdot Q\left(t\right)\right)e^{j\omega_1 t}\right) \\ &\quad = Re\left(\left(I\left(t\right)+j\cdot Q\left(t\right)\right)\cdot e^{j\cdot(\omega_1-\omega_2)\cdot t}\cdot e^{j\omega_2 t}\right) \end{aligned} \tag{2.5}$$

Based on (2.5) a shift map for the baseband representation can be defined

$$\begin{aligned} &shift_{BB}: X_{BB}(f_1) \to X_{BB}(f_2) \\ &\text{with} \\ &\left(I\left(t\right), Q\left(t\right), t\right) \mapsto \\ &(I(t)\cdot\cos\psi(t) - Q(t)\cdot\sin\psi(t), I(t)\cdot\sin\psi(t) + Q(t)\cdot\cos\psi(t), f_2) \\ &\text{and} \\ &\psi(t) = (\omega_1-\omega_2)\cdot t \text{ or} \\ &\psi(t) = (\omega_1-\omega_2)\cdot t \bmod 2\pi \end{aligned} \tag{2.6}$$

This map should only be applied in baseband modeling for small $\frac{(\omega_1-\omega_2)}{\omega_1}$ (see also [7]). Shifting can be interpreted as a map with the describing function $N(A,f) = \cos\psi(t) + j\cdot\sin\psi(t)$.

Differentiation

It is evident that the time domain differential operator D_{PB} in the passband is given by

$$\begin{aligned} &D_{PB}: X_{PB}(f) \to X_{PB}(f) \\ &\text{with} \\ &I(t)\cdot\cos(\omega t) - Q(t)\cdot\sin(\omega t) \mapsto \\ &(I'(t) - \omega\cdot Q(t))\cos(\omega t) - (Q'(t) + \omega\cdot I(t))\sin(\omega t) \end{aligned} \tag{2.7}$$

where $I'(t)$ and $Q'(t)$ are the time derivatives of $I(t)$ and $Q(t)$ resp. ($\omega = 2\pi f$). Thus, the associated differential operator D_{BB} in the baseband is

$$\begin{aligned} &D_{BB}: X_{BB}(f) \to X_{BB}(f) \\ &\text{with} \\ &(I(t), Q(t), f) \mapsto \\ &\left((I'(t) - \omega\cdot Q(t)), (Q'(t) + \omega\cdot I(t)), f\right) \end{aligned} \tag{2.8}$$

If only the steady state is of interest, the time derivatives $I'(t)$ and $Q'(t)$ can be assumed to be 0. In this case, D_{BB} is characterized by the describing function $n(A,f) = j\cdot\omega$. (2.8) is used in [10] to derive baseband descriptions of basic elements as capacitances and inductances. The function can also be used to express general linear transfer functionality in the baseband as will be shown in the following example.

Example 3
A Laplace transfer function is given by

$$H(s) = \frac{b_0 + b_1 \cdot s + \cdots b_k \cdot s^k}{a_0 + a_1 \cdot s + \cdots a_n \cdot s^n} = \frac{Y(s)}{U(s)} \qquad (k < n, a_n \neq 0)$$

where s is the Laplace operator. It is well known that a system of ordinary differential equations represents the same behavior in the time domain (see, e.g. [4]). This system of equations describes the behavior using the passband approach

$$\begin{aligned} x'(t) &= A \cdot x(t) + B \cdot u(t) \\ y(t) &= C \cdot x(t) \end{aligned} \qquad (2.9)$$

with

$$x(t) = (x_0(t), x_1(t), \cdots x_{n-2}(t), x_{n-1}(t))^T \in \mathbb{R}^n$$

$$A = \begin{pmatrix} 0 & 1 & 0 & \cdots & 0 \\ 0 & 0 & 1 & \cdots & 0 \\ \vdots & & \ddots & \vdots & \vdots \\ 0 & 0 & 0 & \cdots & 1 \\ -\frac{a_0}{a_n} & -\frac{a_1}{a_n} & -\frac{a_2}{a_n} & \cdots & -\frac{a_{n-1}}{a_n} \end{pmatrix}, \quad B = \begin{pmatrix} 0 \\ 0 \\ \vdots \\ 1 \end{pmatrix}$$

$$C = (b_0, b_1, \ldots, b_k, 0, \ldots, 0)$$

Considering (2.8), the equivalent baseband description can easily be derived. The structure of the system of equations is given by

$$\begin{pmatrix} x_I'(t) \\ x_Q'(t) \end{pmatrix} = \begin{pmatrix} A & \omega \cdot E \\ -\omega \cdot E & A \end{pmatrix} \cdot \begin{pmatrix} x_I(t) \\ x_Q(t) \end{pmatrix} + \begin{pmatrix} B \cdot u_I(t) \\ B \cdot u_Q(t) \end{pmatrix} \qquad (2.10)$$

$$\begin{pmatrix} y_I(t) \\ y_Q(t) \end{pmatrix} = \begin{pmatrix} C & 0 \\ 0 & C \end{pmatrix} \cdot \begin{pmatrix} x_I(t) \\ x_Q(t) \end{pmatrix} \qquad (2.11)$$

Figure 2.2 shows a bit stream modulated by a 1.575 GHz sinusoidal waveform that is filtered by a bandpass. The magnitude of the output baseband waveform represents the envelope of the passband output waveform as expected.

The main consequences of the considerations in this section is that the baseband representation of a waveform should combine I- and Q-components and the carrier frequency. It seems that it is not necessary to save the phase as proposed in [7]. In case of different carrier frequencies, the operators used in baseband descriptions must define how to handle shifting of the carrier frequency. Last but not least, it should be mentioned that the strong application of the results from the describing function theory simplifies baseband modeling.

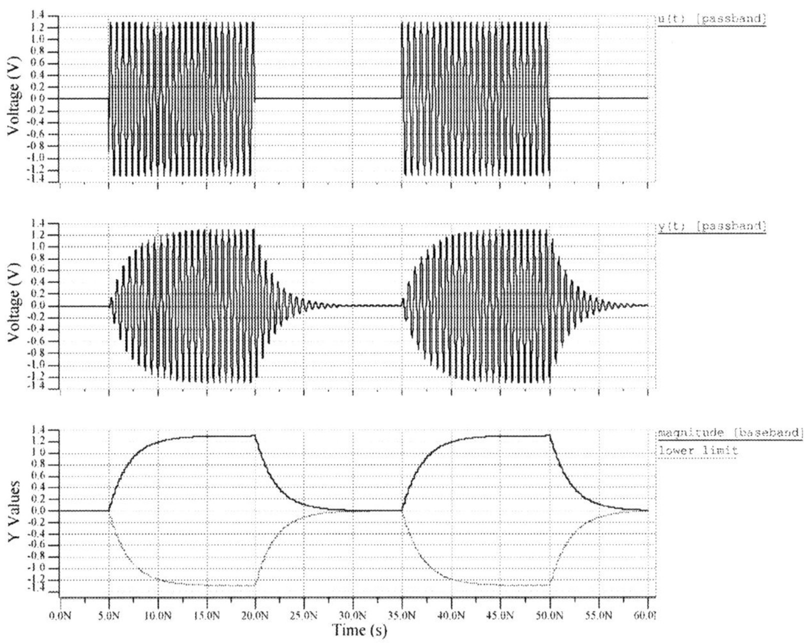

Figure 2.2. Input $u(t)$, passband output $y(t)$, and baseband waveform.

3. Baseband Modeling Using VHDL-AMS

The connection point in the baseband description should carry the information about in-phase, quadrature component, and carrier frequency. To handle this, nature BB_NATURE_VECTOR is declared as follows:

```
nature BB_NATURE is
   REAL   through
   BB_RER reference;
nature BB_NATURE_VECTOR is
   array (NATURAL range <>)
   of BB_NATURE;
```

A multidimensional connection point T in a baseband model can then be declared by

```
terminal  T : BB_NATURE_VECTOR (1 to 2);
signal    F : REAL;
```

When the dependency of transfer characteristics with respect to the carrier frequency shall is analyzed, a quantity port instead of the signal port should be

Table 2.1. Operators for passband and baseband descriptions.

Passband	Baseband
$x : \mathbb{R}^{+}+ \to \mathbb{R}$ (see (2.1))	$X = I + j \cdot Q$ represents x
$x_1 + x_2$	$X_1 + X_2$
$x_1 - x_2$	$X_1 - X_2$
$h\left(x, \frac{dx}{dt}\right)$	$N(A, f) \cdot X$
	where $N : \mathbb{R} \times \mathbb{R} \to \mathbb{C}$ is the describing
	function associated with h (see ([5])

used. Furthermore, we assume that the constitutive relations of the branches in the passband description can be expressed in the following manner

$$g(i, v, s) \text{ with } g : \mathbb{R}^n \times \mathbb{R}^m \times \mathbb{R}^p \to \mathbb{R} \tag{2.12}$$

where i, v, and s are vectors that combine branch voltages and currents and free quantities. It should be possible to express g using the operators given in Table 2.1. The same carrier frequency is used for all operands.

To assign a baseband model to a passband model the following main steps have to be carried out:

- Replace `ELECTRICAL` terminals by connection points as described by the code example for the multidimensional connection point `T`. If a terminal is connected to a voltage or current source branch the signal port is of mode out.

- Instead of real scalar free quantities in the passband description two-dimensional real vectors that carry the in-phase and quadrature component are declared.

- The passband constitutive relations (2.12) should be transformed to baseband relations using Table 2.1. Arguments are mapped to complex numbers. The baseband operations are carried out in the complex area.

- If necessary a frequency shifting is carried out in the model (see equation (2.6)).

- `'DOT` attributes can be replaced considering (2.7) and (2.8).

Auxiliary functions and constants that support the conversion between different types and representations can be summarized as well as the baseband nature descriptions in an additional package.

Example 4

The main part of the passband model of a low noise amplifier (see Fig. 2.3) is given by

```
entity LNA is
 generic (
  A1, A3, RIN, ROUT : REAL);
 port (
  terminal N1, N2 :  ELECTRICAL);
end entity LNA;
```

```
architecture BASIC of LNA is
 quantity VIN  across IIN  through N1;
 quantity VOUT across IOUT through N2;
begin
 0.0 == VIN  - RIN*IIN;
 0.0 == VOUT - ROUT*IOUT - MPB(VIN);
end architecture BASIC;
```

The main parts of the associated baseband model are

```
entity LNA is
 generic (
  A1, A3, RIN, ROUT : REAL);
 port (
 terminal N1:BB_NATURE_VECTOR(1 to 2);
 signal   F1:REAL;
 terminal N2:BB_NATURE_VECTOR(1 to 2);
 signal   F2:out REAL);
end entity LNA;
```

```
architecture BASIC of LNA is
  ...
  function B_OUT(
    I      :BB_NATURE_VECTOR'THROUGH;
    V      :BB_NATURE_VECTOR'ACROSS;
    VCTRL :BB_NATURE_VECTOR'ACROSS)
  return REAL_VECTOR is
    variable RESULT : COMPLEX;
  begin
    RESULT
      := CMPLX(V) - ROUT*CMPLX(I)-
         DF(ABS(VCTRL))*CMPLX(VCTRL);
  return (RESULT.RE, RESULT.IM);
  end function B_OUT;
  quantity VIN  across IIN  through N1;
  quantity VOUT across IOUT through N2;
begin
  ZERO == B_IN(IIN, VIN);
  ZERO == B_OUT(IOUT, VOUT, VIN);
  F2   <= F1;
end architecture BASIC;
```

`ZERO` is a constant `(0.0, 0.0)` of type `REAL_VECTOR (1 to 2)`. The type `COMPLEX` is declared in the package `MATH_COMPLEX` of the `IEEE` library. In this package, overloaded functions for $+$ and $-$ in the complex area and multiplication with a real constant are also declared. `DF` is the complex describing function associated with $N(A, f)$ (see example 1). `ABS` and `CMPLX` are overloaded

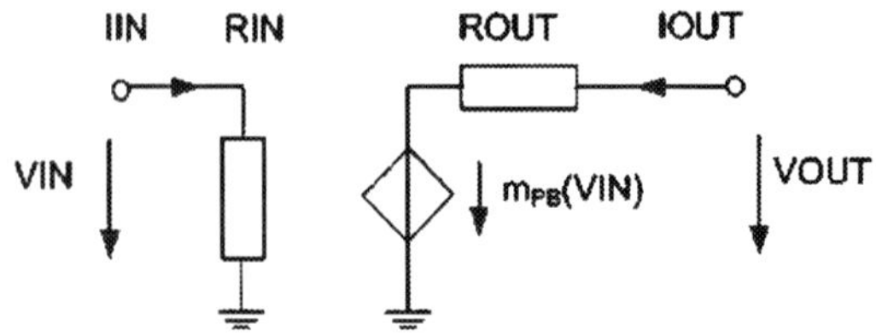

Figure 2.3. Structure of a LNA model.

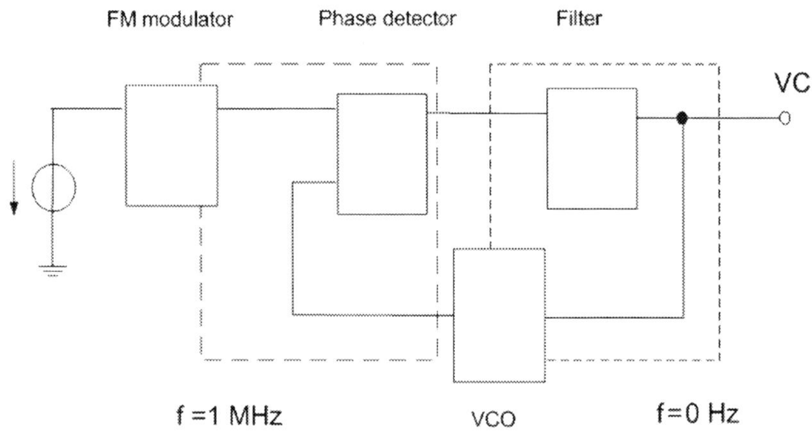

Figure 2.4. Analog PLL.

functions of complex maps. The passband model and the baseband model are compiled into different resource libraries. The structural descriptions that use these models have to be modified with respect to the different connection points.

4. Further Illustrative Examples

Analog PLL

The FM modulator (see Fig. 2.4) is controlled by an electrical voltage source. The carrier frequency of the FM signal is 1 MHz. This is also the center frequency of the VCO. The phase detector is a multiplier that realizes a down-conversion to f = 0 Hz.

Figure 2.5 shows the demodulated waveform VC using passband and baseband models. The example demonstrates the usage of different carrier frequencies in a baseband description.

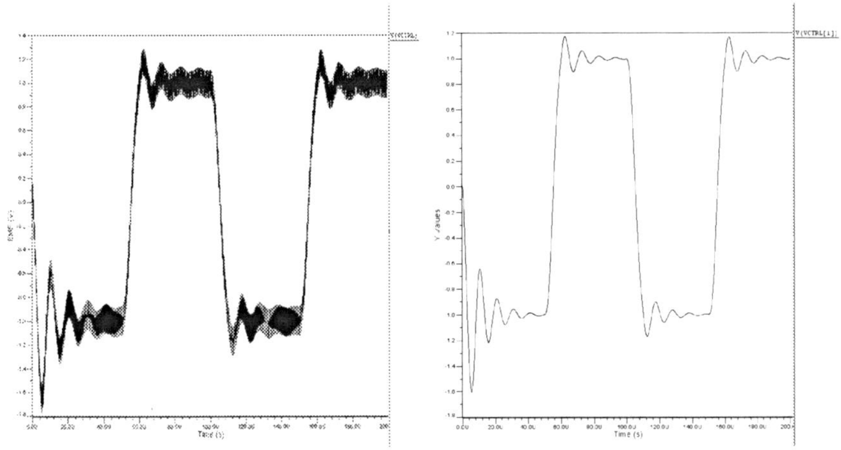

Figure 2.5. Voltages VC of passband and baseband model.

Exited Duffing Oscillator

An excited Duffing oscillator is described by

$$\frac{d^2x}{dt^2} + 2\gamma \cdot \frac{dx}{dt} + \omega_0^2 \cdot x + \beta \cdot x = k \cdot \cos(\omega t) \tag{2.13}$$

Then, the associated baseband description is

$$DF_{DUFF}\left(\sqrt{I^2 + Q^2}, \omega\right) \cdot (I + j \cdot Q) = k \tag{2.14}$$

with $DF_{DUFF}(A, \omega) = -\omega^2 + j \cdot 2\gamma\omega + \omega_0^2 + \frac{3}{4} \cdot A^2$

It is known that for some ω more than one solution (I, Q) of (2.14) can be determined [2]. That means, there may exist multiple steady state solutions of (2.13). Combining (2.13) with a curve tracing algorithm these characteristics can be determined [11, 6]. Using the describing function, the basic equation (2.14) to determine the frequency response can easily be established. Figure 2.6 shows the magnitude for $\gamma = 0.05, \omega_0^2 = 1, \beta = 1, k = 0.2$. In this example the carrier frequency is not fixed.

5. Summary

The verification of front ends for digital communication requires efficient modeling and simulation methods. In order to be able to compare different system architectures a high execution speed of the simulation is required. One possibility to reduce simulation time is a higher level of abstraction of the analog models. Baseband models can be used for this purpose. Other approaches are summarized in [9, 12].

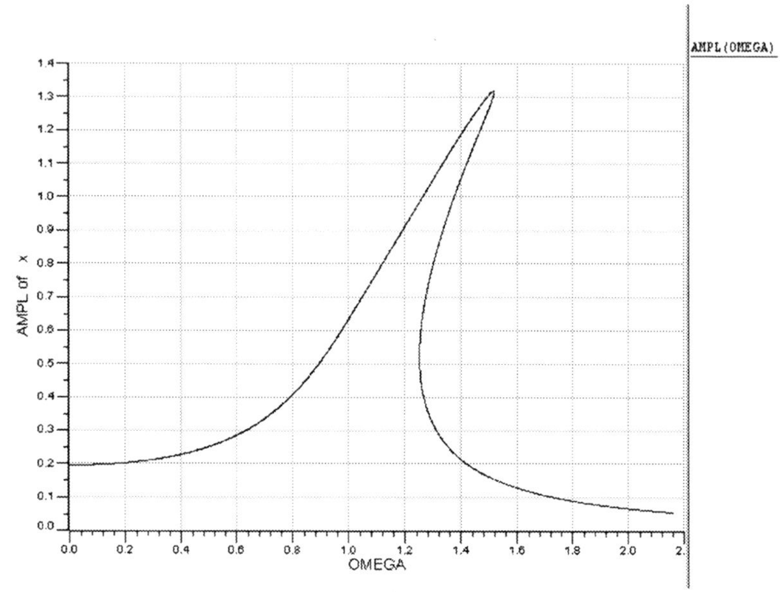

Figure 2.6. Magnitude of Duffing oscillator.

Baseband models are a low-pass waveform representation of passband waveforms. The passband waveforms are characterized by a narrow frequency spectrum around a carrier frequency. With this baseband representation, in-band distortions for nonlinear blocks can be investigated, for instance.
This chapter formally defines a relation between passband and equivalent baseband models. Thus, it formally describes how to write consistent passband and baseband models. This approach can easily be implemented using behavioral description languages. It is shown how the approach works using VHDL-AMS. Some implementation details were discussed. The approach can be used with other languages and description methods as for instance Verilog-AMS or SystemC-AMS in a similar way.

The main passband functionality can be expressed using linear transfer functions and static nonlinearities. There exists a simple relation between linear transfer functions in the passband and baseband representations. Using describing functions for baseband modeling helps to establish consistent baseband and passband descriptions of the nonlinear parts. This approach makes it possible to access the former results from the describing function theory. The idea is formally described as Modeling procedure in Section 2. Example 1 shows how it works.

A special waveform representation that also considers the frequency makes it possible to switch the carrier frequency between different parts of a system. The carrier frequency is usually constant in a part of a system. Thus in VHDL-AMS, it is represented by a time-discrete signal.

References

[1] IEEE Standard VHDL Analog and Mixed-Signal Extensions. IEEE Std 1076.1-1999. Design Automation Standards Committee of the IEEE Computer Society, March 18, 1999.

[2] Brandt, S. and Dahmen, H.D. *Mechanik*. Berlin: Springer, 1996.

[3] Chen, J.E. Modeling RF Systems. March 6, 2005. Available: http://www.designers-guide.org/Modeling/

[4] Geering, H.P. *Regelungstechnik*. Berlin: Springer, 2001.

[5] Gelb, A. and Vander Velde, W.E. *Multiple-input Describing Functions and Nonlinear System Design*. New York: McGraw-Hill, 1968. Available: http://en.wikipedia.org/wiki/Describing_function

[6] Haase, J., Pönisch, G. and Uhle, M. Multiple DC solution determination using VHDL-AMS. Proc. BMAS 2003, October 2003, pp. 107–112. Available: http://www.bmas-conf.org/2003

[7] Joeres, S. and Heinen, S. Mixed-mode and mixed-domain modelling and verification of radio frequency subsystems for SoC applications. Proc. BMAS 2005, September 2005, pp. 54–59. Available: http://www.bmas-conf.org/2005

[8] Kundert, K. Future directions in mixed-signal behavioral modeling. Keynote BMAS 2002, October 2002. Available: http://www.bmas-conf.org/2002

[9] Root, D., Wood, J. and Tufillaro, N. New techniques for nonlinear behavioral modeling of microwave/RF ICs from simulation and nonlinear microwave measurement. Proc. DAC 2003. June 2003, pp. 85–90.

[10] Tuinenga, P.W. Models rush in where simulators fear to tread: extending to the baseband-equivalent method. Proc. BMAS 2002, October 2002, pp. 31–40. Available: http://www.bmas-conf.org/2002

[11] Ushida, A., Yamagami, Y. and Nishio, Y. Frequency response of nonlinear networks using curve tracing algorithm. Proc. ISCAS 2002, pp. 641–644.

[12] Wambacq, P., Vandersteen, G., Rolain, Y., Dobrovolny, P., Goffioul, M and Donnay, S. Dataflow simulation of mixed-signal communication circuits using a local multirate, multicarrier signal representation. IEEE Trans. Circ. Syst I vol. 49, pp. 1554–1562, November 2003.

Chapter 3

VERIFICATION-ORIENTED BEHAVIORAL MODELING OF NONLINEAR ANALOG PARTS OF MIXED-SIGNAL CIRCUITS

Martin Freibothe[1], Jens Döge[1], Torsten Coym[1], Stefan Ludwig[1], Bernd Straube[1], and Ernst Kock[2]

[1]*Fraunhofer-Institut für Integrierte Schaltungen IIS*
Branch Lab Design Automation
Zeunerstr. 38, 01069 Dresden
{martin.freibothe,jens.doege,torsten.coym,stefan.ludwig,bernd.straube}@eas.iis.fraunhofer.de

[2]*Infineon Technologies AG*
Neubiberg
ernst.kock@infineon.com

Abstract In this work, an approach to the "verification-oriented" modeling of the analog parts' behavior of mixed-signal circuits is presented. Starting from a continuous-time, continuous-valued behavioral representation of an analog part in terms of a differential-algebraic equation system, a discrete-time, discrete-valued behavioral model is derived. This kind of model both captures dynamic aspects of the analog behavior and can be implemented using the synthesizable subset of a hardware description language like VHDL. With the help of the proposed approach, the continuous-time, continuous-valued analog parts' behavioral descriptions can be replaced by digital behavioral models leading to a verification-oriented model of the underlying mixed-signal circuit. The resulting model can be formally verified using established methods and tools from formal digital verification.

Keywords Behavioral modeling, mixed-signal verification, semiformal methods

A.S. Huss (ed.), Advances in Design and Specification Languages for Embedded Systems – Selected Contributions from FDL'06, 37–51.

1. Introduction

In today's circuit designs, the integration of analog parts on system-on-chips (SoCs) and ASICs is common practice. Usually in such designs the analog parts comprise a comparatively small fraction of the overall chip area. On the contrary, a large part of the development time of a mixed-signal circuit is spent on the analog parts.

Using well-established formal verification techniques, digital circuit designs can be formally verified in order to find design errors. In contrast to simulation-based verification, these techniques reach a 100% coverage of the functionality of a circuit design. Currently there are no tools allowing for the formal verification of analog and mixed-signal circuits. In today's verification flows the analog and digital parts are verified separately by simulation and formal verification techniques, respectively. The verification of the overall behavior of a mixed-signal circuit and, hence, the connection of the analog and digital parts is then carried out based on simulation. All simulation-based verification methods have the disadvantage of requiring a testbench and simulation stimuli. Despite the high manual effort to provide both of them, due to the complexity of today's designs only a relatively low coverage of a circuit's functionality can be achieved.

To simplify matters, those parts of a mixed-signal circuit that are represented by continuous-time, continuous-valued behavioral models are in the following referred to as 'analog components'. An approach to the verification-oriented modeling of analog components is presented. Starting from a behavioral description of an analog component in terms of nonlinear differential-algebraic equations a digital model is derived by discretizing the time and quantizing the continuous values. Based on the resulting digital behavioral model some properties of practical relevance can be proved using commercially available *bounded model checking* tools. The presented work extends the class of mixed-signal circuits that can be verified with the help of formal methods that are commonly used for the verification of digital circuits.

The remainder of the paper is organized as follows: In Section 2 related work is reviewed. Section 3 describes the proposed modeling approach and the application of formal verification techniques. An example of the application of the proposed flow is given in Section 4 and some concluding remarks are presented in Section 5.

2. State of the Art

Simple mixed-signal circuits can be represented by hybrid automata [1, 10]. The verification of a circuit design is carried out by performing a reachability analysis over the state-space of the corresponding hybrid automaton. Even under strong restrictions, such as piecewise constant derivatives of the variables, the reachability problem remains undecidable for this class of automata, for the state-space is infinite in general [11]. Recent approaches to the verification of hybrid automata utilize different techniques to approximate the reachable state sets. Standard model checking algorithms are then applied to these finite-state approximations [8, 9]. The approach proposed in [3] suggests to perform a reachability analysis on a certain class of hybrid systems that can be represented as mixed logical dynamical or piecewise affine systems to get a piecewise linear approximation of the exact solution. Drawbacks of these approaches are that they introduce an approximation error, are not sound [9], or their performance strongly depends on the shape of the reachable sets of states [3].

An approach to the verification of linear analog circuits is presented in [2]. The circuit's specification is given in terms of its transfer function. Based on the state equations, the transfer function of the implementation of the circuit is derived. Both transfer functions are transformed into the discrete-time domain using Z-transform. The continuous values denoting voltages and currents are quantized. The resulting discrete-time, discrete-valued transfer functions of both the specification and the implementation can be represented by finite deterministic automata, respectively. The verification is then carried out by an equivalence-check, an approach to property-checking is not mentioned in this work. The proposed verification flow is restricted to linear analog circuits. The problem of arithmetic overflows that might occur during verification is mentioned; however, overflows are not recognized or treated by this approach.

In [12] an approach to the semiformal verification of the static behavior of mixed-signal circuits is presented. The proposed approach is based on SAT-based property-checking. Starting point of the described verification flow is the static behavioral description of a mixed-signal circuit in VHDL. In such a circuit description, VHDL's floating point type *real* is used to characterize the analog behavior. Given such a mixed-signal circuit description, a verification-oriented model is derived for which SAT-based property-checking is carried out. In the verification-oriented model the floating point type *real* used to describe analog behavior is approximated by integer numbers of a finite interval. Such a model can be represented by a finite Mealy automaton and thus be verified using well-established formal verification methods. Possible overflows due to arithmetic operations are recognized during the verification process and cause the corresponding property to fail. The quantization error that is introduced when the verification-oriented model is derived has to be considered in the properties to be verified.

A similar approach to the verification of the static behavior of mixed-signal circuits is proposed in [6]. In this approach the rational numbers $\mathbb{Q}$ are used to represent analog values. The mixed-signal circuit is given as an infinite automaton that is described in a simple XML. A validity checker that allows the usage of arithmetic expressions in the formulae is utilized to prove properties based on the automaton. The accuracy of the internal representation of the rational numbers is only limited by the amount of memory that is available so that arithmetic overflows do not occur in practice.

Both of the last two mentioned approaches use a behavioral description of a mixed-signal circuit that abstracts from the dynamic analog behavior to static, piecewise linear behavior with respect to the clock-cycle of the digital part of the mixed-signal circuit. Neither of these two approaches describes how such a static behavioral model of an analog component can be derived.

Based on [9], the authors of [7] propose an approach to the verification of nonlinear analog circuits. The state-space of an analog circuit that is given by the circuit's inputs and independent state-variables is restricted to a finite region that is approximated using hyperboxes. Discrete state-transitions of randomly chosen representatives of the discrete states are determined by over-approximating the solutions of differential-algebraic equations. Additionally, these transitions are annotated with timing information so that timing constraints can also be verified. Due to complexity issues the presented approach is only applicable for comparatively small analog circuits. The verification of mixed-signal circuits is not considered in this approach.

3. Verification-Oriented Modeling and Verification

This work proposes to model the behavior of the analog components of a mixed-signal circuit with the help of finite automata. The different modeling steps, starting from the behavioral description of an analog component in terms of a differential-algebraic equation system, are depicted in Fig. 3.1.

The result of the proposed modeling flow is a digital behavioral model ("single-step automaton"). Using this representation some practically relevant properties of an analog component can be semiformally verified by bounded model checking through property-checking using the approach presented in [12]. Arithmetic overflows that might occur due to an inappropriate quantization of the continuous values are recognized during verification and appropriately treated so that no *false-positives* occur because of overflows. Based on a single-step automaton another, more abstract model ("n-step automaton") can be derived. By replacing the continuous-time, continuous-valued behavioral

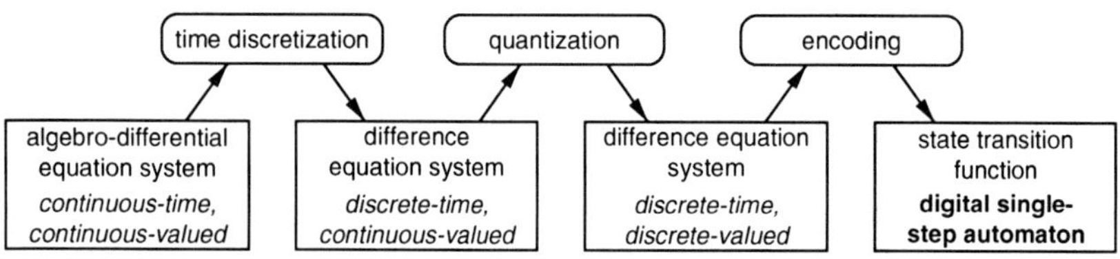

Figure 3.1. Proposed modeling flow.

models of the analog components by their corresponding n-step automata in the interconnection with the digital parts, a verification-oriented model of the whole underlying mixed-signal circuit can be created.

Discretization and Quantization

The presented approach to the derivation of verification-oriented models of analog components is illustrated using the example circuit shown in Fig. 3.2. Without loss of generality a linear, single-input, single-output analog component has been chosen for illustrating the approach.

Based on the circuit's netlist a behavioral model in terms of a differential-algebraic equation system is derived. Such an equation system is the starting point for the proposed modeling flow. The behavior of the circuit depicted in Fig. 3.2 is described by a first-order linear differential equation system:

$$\begin{bmatrix} \dot{I}(t) \\ \dot{V}_{\text{out}}(t) \end{bmatrix} = \begin{bmatrix} -\frac{R}{L} & -\frac{1}{L} \\ \frac{1}{C} & 0 \end{bmatrix} \cdot \begin{bmatrix} I(t) \\ V_{\text{out}}(t) \end{bmatrix} + \begin{bmatrix} \frac{1}{L} \\ 0 \end{bmatrix} \cdot V_{\text{in}}(t). \tag{3.1}$$

The state equation system (3.1) can be solved analytically, however, in general this is not the case, especially for practically relevant analog components. Due to that fact such a behavioral description of analog components is not suitable for the practical application of formal verification techniques. The result of the proposed approach to verification-oriented modeling is a behavioral description of an analog component that both represents dynamic analog behavior and can be verified using the well-established formal tools from the verification of digital circuits.

Based on the behavioral description in terms of a differential-algebraic equation system, in a first step the time is discretized (cf. Fig. 3.1). Linear differential equation systems like (3.1) can be time-discretized using Z-transform. In general, particularly regarding nonlinear analog components, different numerical integration methods like, e.g. implicit or explicit Euler or the trapezoidal rule

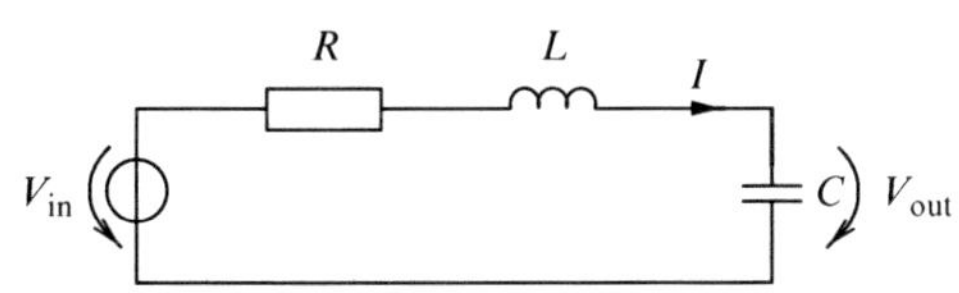

Figure 3.2. RLC circuit.

can be applied. The difference equation system (3.2) results from (3.1) by applying the trapezoidal rule:

$$\begin{aligned}\begin{bmatrix} I(k+1) \\ V_{\text{out}}(k+1) \end{bmatrix} = \frac{1}{K_1} \cdot & \left[\begin{bmatrix} K_2 & -\frac{1}{L} \cdot h \\ \frac{1}{C} \cdot h & K_3 \end{bmatrix} \cdot \begin{bmatrix} I(k) \\ V_{\text{out}}(k) \end{bmatrix} \right. \\ & \left. + \begin{bmatrix} \frac{1}{L} \cdot \frac{h}{2} \\ \frac{1}{L \cdot C} \cdot \frac{h^2}{4} \end{bmatrix} \cdot (V_{\text{in}}(k) + V_{\text{in}}(k+1)) \right], \quad k \in \mathbb{Z}\end{aligned} \tag{3.2}$$

with the constants:

$$K_1 = 1 + \frac{R}{L} \cdot \frac{h}{2} + \frac{1}{L \cdot C} \cdot \frac{h^2}{4}, \tag{3.3}$$

$$K_2 = 1 - \frac{R}{L} \cdot \frac{h}{2} - \frac{1}{L \cdot C} \cdot \frac{h^2}{4}, \tag{3.4}$$

$$K_3 = 1 + \frac{R}{L} \cdot \frac{h}{2} - \frac{1}{L \cdot C} \cdot \frac{h^2}{4}. \tag{3.5}$$

In these formulae the constant $h \in \mathbb{Q}^+$ denotes the time step used for the numerical integration, i.e. the difference between two consecutive points in time that are represented by the sequences that determine V_{out} and I, respectively. In order to be able to represent the discrete-time behavioral description of the example circuit (cf. Fig. 3.2) given by (3.2) with the help of a finite automaton, the continuous-valued currents and voltages have to be quantized and restricted to finite ranges of values, respectively. This step is necessary to get finite sets of input, state, and output symbols for the automaton. The resolution of the quantization and the finite intervals' boundaries are determined by the environment the analog component is used in.

The choice of these parameters has an impact on the complexity of the resulting automaton and, hence, on the complexity of the subsequent verification task. Both the resolution of the quantization and the time step h for the numerical integration have to be appropriately chosen to get an automaton that sufficiently reflects relevant aspects of the underlying analog behavior. According to a class of numerical integration techniques the automaton representing the analog behavior for one time step h is referred to as "single-step automaton" (cf. Fig. 3.1).

Building a Single-Step Automaton

Any possible combination of input values and initial states of an analog component has to be taken into account when building a single-step automaton to represent the analog behavior. The state-variables of an analog component are in general represented by currents through inductors and voltages across capacitors. Considering the example circuit depicted in Fig. 3.2 there are two state-variables: the current I through the inductor L and the voltage V_{out} across the capacitor C. The voltage V_{out} also denotes the output variable of the circuit.

According to the behavioral description of the circuit in terms of the difference equation system (3.2), the current I and the voltage V_{out} at the time-point denoted by $k+1$, $k \in \mathbb{Z}$, depend on the values of the input voltage at the same time-point and the previous one. To simplify matters, four different sequences of the input voltage V_{in} of length two have been chosen to build the single-step automaton. Any sequence of $0.0\,\text{V}$ and $3.2\,\text{V}$ over two time-points has been considered. Using these values, V_{in} could, for example, represent a digital control input with $0.0\,\text{V}$ denoting logical "0" and $3.2\,\text{V}$ denoting logical "1".

The following constant device parameters have been used to build the single-step automaton for the example circuit: $R = 1\,\text{k}\Omega, L = 1\,\mu\text{H}, C = 0.8\,\text{pF}$. The voltages are quantized with a resolution of $0.4\,\text{V}/\text{bit}$ over the finite interval $-6.4\,\text{V} \ldots 6.0\,\text{V}$, the currents are quantized with $1.25 \cdot 10^{-1}\ \text{mA}/\text{bit}$ between $-2.0\,\text{mA}$ and $1.875\,\text{mA}$. This way, the variables representing voltages or currents can be implemented in an HDL using bit vectors of width five. The time step has been set to $h = 0.7\,\text{ns}$.

The difference equation system (3.2) must be solved for any possible combination of the four mentioned input sequences and any possible initial assignment of the circuit's state-variables I and V_{out}. The results are rounded and mapped to their appropriate quantization interval. With the parameters chosen as aforementioned, there are altogether $4 \cdot 2^5 \cdot 2^5 = 4096$ different combinations of input sequences and initial states the difference equation systems has to be solved for. The result of this step is the discrete-time, discrete-valued representation of the analog behavior as a table of values. Figure 3.3 shows an extract of the example circuit's table of values. Using an appropriate encoding, the real-valued currents and voltages within finite ranges can be represented by

$V_{\text{in}}(k)$	$V_{\text{in}}(k+1)$	$I(k)$	$V_{\text{out}}(k)$	$I(k+1)$	$V_{\text{out}}(k+1)$
0.0	0.0	-1.250	0.8	-0.750	0.0
0.0	0.0	-1.250	1.2	-1.000	0.4
0.0	0.0	-1.375	2.8	-1.750	1.6
	$\vdots$	$\vdots$	$\vdots$	$\vdots$	$\vdots$
3.2	3.2	-1.375	2.4	-0.125	1.6
3.2	3.2	-1.875	2.8	-0.375	1.6
3.2	3.2	-1.750	3.2	-0.625	2.0

Figure 3.3. State-transition function of the example circuit from Fig. 3.2 as a table of values, V in [V] and I in [mA].

bit vectors. The resulting digital representation embodies the state-transition function of the single-step automaton. Within such an automaton the time is represented by the automaton's implicit clock whose period corresponds to the time step h that was chosen for the numerical integration.

With the help of the presented flow the digital behavioral representation as single-step automaton of an arbitrary nonlinear analog component for arbitrary input sequences can be derived. The computational effort of the approach grows exponentially with the number of inputs, state-variables, and the number of quantization steps. It is also possible to utilize a netlist simulator in conjunction with *initial conditions* to get the state-transition function in terms of a table of values.

A state-transition function represented as a table of values can easily be implemented in a hardware description language. For the verification described in the following, the example circuit's table of values (cf. Fig. 3.3) has been implemented in VHDL using the approach presented in [12].

To the best of our knowledge there is currently no way of formally proving the correctness of the derivation of the digital behavioral model from a behavioral description in terms of a differential-algebraic equation system as described above. For that reason whether the abstraction performed to derive the digital model, i.e. the choice of the modeling parameters like the time step h and the quantization parameters has been feasible is validated by simulating/executing both behavioral models for some input sequences and comparing the corresponding traces. Though the derived digital model is not meant to be used for simulation but formal verification, the VHDL implementation is simulated for validation purposes. For this validation task, the initial behavioral description as differential-algebraic equation system can, for example, be implemented and simulated using VHDL-AMS. As expected, there is a significant speedup when the single-step automaton is simulated compared to simulating the differential-algebraic equation system for the same input sequences.

Verification of a Single-Step Automaton

The VHDL implementation of a single-step automaton can be formally verified using well-established *bounded model checking* techniques, for example, with the 360 MV (formerly GateProp) tool [4] from OneSpin Solutions GmbH. With the help of this tool formal properties describing the expected behavior over finite intervals of time are checked for the single-step automaton representing the dynamic behavior of the analog component. However, some restrictions have to be made. A single-step automaton is only valid for input values that have been considered when the automaton was built. Input values that are not valid in this sense are recognized by the model and cause a property to fail.

```
property steady_state is
assume:
 during[t, t+1]:
      either
        V_in = VDD_0V0; or   -- constant denoting 0.0V
        V_in = VDD_3V2;      -- constant denoting 3.2V
      end either;
 during[t+2, t+14]: V_in = prev( V_in );
prove:
 at t+14:  V_out = prev( V_out );
 at t+14:  I     = prev( I );
end property;
```

Figure 3.4. Example *VHI* property for the transient response of the RLC circuit.

Based on a single-step automaton, some properties about the dynamic behavior of analog components can be verified. An example property could state that the output voltage does not fall below or exceed certain values.

A formal property (cf. Fig. 3.4) written in 360 MV's property specification language *VHI* consists of two main parts: a (possibly empty) set of assumptions and a set of commitments. Typically, the assumptions are used to express restrictions on the values at the inputs or the state-variables. The commitments describe the behavior the circuit must exhibit under these restrictions to be correct, mainly in terms of expected values at the outputs and input–output relations. In the assumptions as well as in the commitments it is possible to reference to the value of any signal at any time-point of the trace. A property holds iff the implication *assumptions* $\Longrightarrow$ *commitments* holds for each trace of the circuit starting from any arbitrary state and allowing arbitrary values at the inputs at any time-point.

With the property shown in Fig. 3.4 could be proved, that independent of its initial state, the example RLC circuit settles to a steady-state after 12 clock-cycles of the single-step automaton. Here it is assumed that the steady-state is reached iff both the state-variable denoted by the current I and the output voltage V_{out} do not change provided the input voltage V_{in} is constant.

Further Applications of Single-Step Automata

Based on a single-step automaton, another abstraction step can be performed. The aim is the derivation of a behavioral model of an analog component that represents its behavior with respect to the clock-cycle of the digital part of the mixed-signal circuit. This kind of model can replace the continuous-time, continuous-valued representation of an analog component in the interconnection of the analog and digital parts of a mixed-signal circuit.

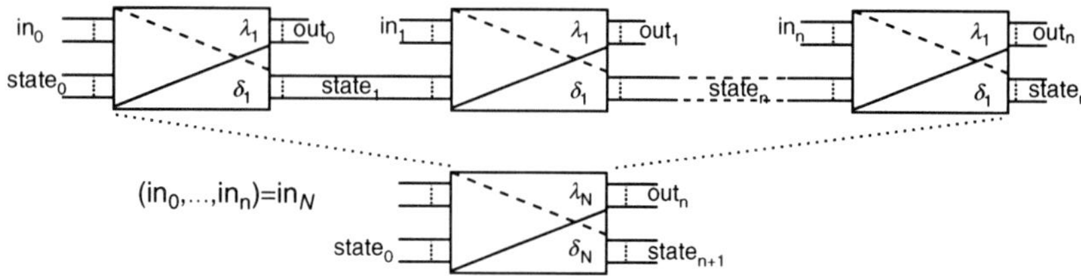

$$\text{out}_n = \lambda_N(\text{state}_0, (\text{in}_0, \ldots, \text{in}_n)) = \lambda_1(\delta_1(\ldots\delta_1(\delta_1(\text{state}_0, \text{in}_0), \text{in}_1), \ldots, \text{in}_{n-1}), \text{in}_n)$$
$$\text{state}_{n+1} = \delta_N(\text{state}_0, (\text{in}_0, \ldots, \text{in}_n)) = \delta_1(\ldots\delta_1(\delta_1(\text{state}_0, \text{in}_0), \text{in}_1), \ldots, \text{in}_{n-1}), \text{in}_n)$$

Figure 3.5. Building an n-step automaton based on a single-step automaton.

Let the time step h used for creating the single-step automaton be chosen such that $h \cdot n = T$, $n \in \mathbb{N}$, with T being the period of the clock-cycle of the mixed-signal circuit's digital part the analog component is connected to. The time step h corresponds to the period of the single-step automaton's implicit clock. The single-step automaton is executed n-times in series as illustrated in Fig. 3.5. The state that is reached after one execution step is the initial state of the consecutive step. Arbitrary but fixed sequences of values in_{N} are chosen for the automaton's inputs (cf. Fig. 3.5) within the bounded interval of time denoted by $[0, n-1]$. The restriction to fixed input sequences is necessary to be able to summarize n consecutive execution steps of the single-step automaton. This way the mapping of the input sequences and initial states to the states and output values after n clock-cycles of the single-step automaton can be summarized in another table of values. This table represents the state-transition function of the corresponding n-step automaton.

An n-step automaton is a digital representation of an analog component's behavior with respect to the clock-cycle of the digital circuitry the analog component is embedded in. Building an n-step automaton abstracts from the dynamic analog behavior in between two capturing edges of the digital clock. In case all output values after n steps of the single-step automaton are independent from the automaton's initial state, the corresponding n-step automaton represents a combinatorial behavior, sequential behavior otherwise.

There are different application scenarios for the verification-oriented representation of a mixed-signal circuit's analog component by a single-step automaton. Well-established formal verification techniques from the field of digital verification can be applied in different ways:

- Formal verification of properties about the transient response of the analog component represented by the single-step automaton.

- Composition of different single-step automata in order to get a verification-oriented model of the corresponding analog component's interconnection. The result of such a composition of single-step automata is again a single-step automaton.

- Derivation of an n-step automaton and replacing the continuous-time, continuous-valued behavioral model in the interconnection with the digital parts of the mixed-signal circuit. The resulting digital behavioral model of the mixed-signal circuit can be formally verified using the approach presented in [12].

- Composition of different n-step automata. The resulting digital behavioral description can be used to replace the continuous-time, continuous-valued behavioral description in order to get a verification-oriented model of a mixed-signal circuit.

In case the characteristic time constant of an analog component is larger than the period of the digital part's clock-cycle, i.e. the digital part is faster compared to the analog part, the time step h must be chosen such that it is a multiple integer of T, that is $h = T \cdot m$, $m \in \mathbb{N}$. For this kind of mixed-signal circuit, the digital model covering the overall behavior has two clock inputs: one is the original clock input of the digital part and another one for the analog part that has a clock period which is equal to h. For the verification of such a model, the utilized *bounded model checker* needs to support some kind of *clocking-scheme* where two clock inputs and their fixed ratio denoted by m can be defined. The 360 MV tool does provide this feature.

Using the presented approach to represent analog behavior with the help of single-step automata and/or n-step automata extends the class of mixed-signal circuits formal verification techniques are applicable to.

4. Example Circuit: Pixel Cell

The method presented in Section 3 has been applied to create a verification-oriented model of the analog component depicted in Fig. 3.6. The figure shows a pixel cell together with an analog switched current (SI) cell. This kind of pixel cell can, for example, be part of an image sensor matrix where an SI cell as a readout circuit is assigned to each column.

The photocurrent I_{NW} through the photodiode D_{NW} of the pixel cell that is excited by a light source is considered as the input of the analog component. For the depicted equivalent network for analyzing the transient behavior, the voltage $V_{C\mathrm{SI}}$ across C_{SI} is the analog component's output. The remaining input voltages denoted by V_{Bias}, V_{Sel}, and V_{Act} are of constant value for the considered configuration.

For the application of the presented approach to verification-oriented modeling the circuit depicted in Fig. 3.6 has been divided into two feedback-free partitions 1 and $2(ab)$. The assumption about these partitions being feedback-free

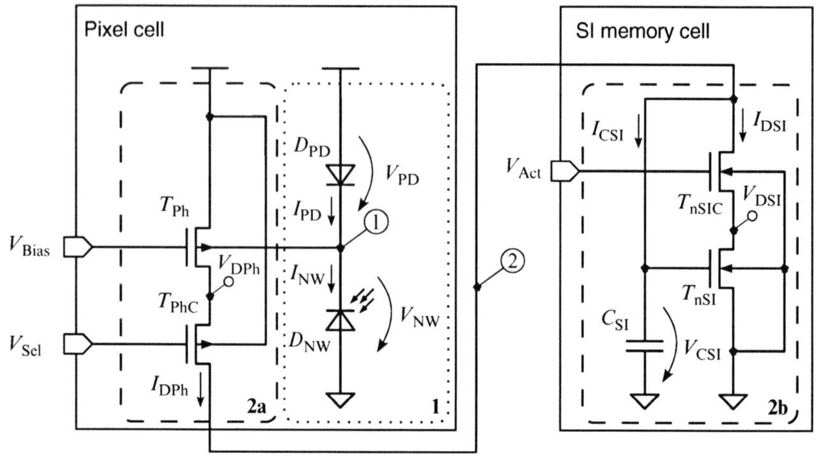

Figure 3.6. Equivalent network of the modeled pixel cell with SI memory cell.

is valid with sufficient accuracy for the application considered here. However, finding a partitioning for an arbitrary circuit that exhibits this property requires careful investigations and a deep understanding of the circuit's behavior. In terms of the derivation of the verification-oriented model of the depicted circuit, the two partitions 1 and $2(ab)$ can be considered separately. The partitioning step is necessary as the dominant time constants of these partitions differ by six orders of magnitude. The time constants determine the time step h so that an approximation of the dynamic analog behavior with reasonable accuracy can only be achieved with the help of two different single-step automata with the time steps h_1 and h_2, respectively. The parameters for the quantization of the continuous values in terms of the resolution and the intervals were chosen equally for both of the single-step automata in order to allow for the later composition of them.

The behavioral descriptions in terms of nonlinear differential-algebraic equation systems for both of the two partitions are derived using modified nodal analysis. For the transistors simplified EKV models [5] are used in this initial behavioral description. In comparison to the original EKV models, the applied EKV models neglect channel length modulation and velocity saturation as these parameters have a negligible impact on the circuit's behavior in the configuration considered here.

The parameters for the photocurrent's (I_{NW}) resolution and the quantization interval are determined. Based on these parameters the single-step automata representing the behavior of partition 1 and partition $2(ab)$, respectively, are built. The output of the automaton representing partition 1 is the voltage at node

① which is in turn the input for the second single-step automaton representing partition $2(ab)$.

The behavior of the analog component depicted in Fig. 3.6 with respect to the digital clock of the surrounding circuitry can be represented by an n-step automaton. Therefore, at first the n_2-step automaton for partition $2(ab)$ that settles six times faster compared to partition 1 is built. For partition 1, the n_1-step automaton is built whereas the number of consecutive executions of the corresponding single-step automaton is equal to the quotient of the period of the digital clock T and the time step h_1; T needs to be an integer multiple of h_1. Finally, the composition of the n_1-step automaton with the n_2-step automaton represents the behavior of the pixel cell with respect to the digital clock.

Due to complexity reasons, it is not feasible to represent the whole relevant value domain for the input photocurrent within one verification-oriented model. Instead, the relevant value domain was covered with the help of several different models each representing the behavior for disjoint intervals of the input current according to the *divide and conquer* principle. Any input values violating the allowed ranges are recognized in the VHDL implementation and cause a property to fail during verification. Using the 360 MV tool, some properties about the transient behavior of the output voltage V_{CSI} of the modeled circuit (cf. Fig. 3.6) could be proved. Proving a single property took about 40 minutes of CPU time on state-of-the-art computer hardware.

5. Conclusion

Based on behavioral descriptions of the analog components of a mixed-signal circuit in terms of – in general nonlinear – differential-algebraic equation systems an approach to verification-oriented modeling has been presented. The result of the modeling process is a digital behavioral model, the so-called single-step automaton. A single-step automaton can be implemented using the synthesizable subset of a digital hardware description language.

A single-step automaton represents an analog component's behavior for a certain time step. Provided that the parameters for the numerical integration and the resolution for the quantization have been chosen appropriately, the single-step automaton represents the dynamic analog behavior with reasonable accuracy. The computational effort of the presented flow grows exponentially with the number of inputs and state-variables of the analog component under consideration and the number of quantization steps.

Based on a single-step automaton, some practically relevant properties about the transient behavior of an analog component can be proved using well-established digital *bounded model checking* tools. In this work, the 360 MV tool from OneSpin Solutions GmbH has been used for property-checking. Despite the utilization of formal verification techniques, due to the discretization of time and the quantization of the electrical quantities the presented flow has to be characterized as *semiformal*. Based on a single-step automaton another, more abstract model of an analog component can be derived that represents the analog behavior with respect to the clock-cycle of the digital part of the underlying mixed-signal circuit. By replacing the continuous-time, continuous-valued behavioral descriptions of all analog components with their representation as n-step automata, a verification-oriented model of the mixed-signal circuit can be derived. Thus, the class of mixed-signal circuits formal verification techniques are applicable to has been extended with the presented approach. In particular, also nonlinear analog components can be modeled and verified by applying this flow.

References

[1] Alur, R., Courcoubetis, C., Henzinger, T. A., and Ho, P.-H. (1993). Hybrid automata: An algorithmic approach to the specification and verification of hybrid systems. In: Grossmann, R. L., Nerode, A., Ravn, A. P., and Rischel, H., editors, *Hybrid Systems*, vol. 736 of *LNCS*, pp. 209–229. Springer.

[2] Balivada, A., Hoskote, Y., and Abraham, J. A. (1995). Verification of transient response of linear analog circuits. In: *13th IEEE VLSI Test Symposium (VTS)*, pp. 42–47. IEEE Computer Society Press.

[3] Bemporad, A., Ferrari-Trecate, G., and Morari, M. (2000). Observability and controllability of piecewise affine and hybrid systems. *IEEE Trans. Automatic Control*, 45:1864–1876.

[4] Bormann, J., Blank, C., and Winkelmann, K. (2005). Technical and managerial data about property checking with complete functional coverage. In: *Proc. Euro DesignCon 2005 (published as CD-ROM)*, Munich, Germany.

[5] Bucher, M., Lallement, C., Enz, C., Théodoloz, F., and Krummenacher, F. (1998). The EPFL-EKV MOSFET model equations for simulation. Technical Report, Electronics Laboratories, Swiss Federal Institute of Technology (EPFL), Lausanne, Switzerland.

[6] Freibothe, M., Schönherr, J., and Straube, B. (2006). Formal verification of the quasi-static behavior of mixed-signal circuits by property checking. In: Maler, O., editor, *Proc. the First Workshop on Formal Verification of Analog Circuits (FAC 2005)*, vol. 153 of *ENTCS*, pp. 23–35, Elsevier, Edinburgh, UK.

[7] Grabowski, D., Platte, D., Hedrich, L., and Barke, E. (2005). Time constrained verification of analog circuits using model-checking algorithms. In: *ENTCS*.

[8] Gupta, S., Krogh, B., and Rutenbar, R. (Nov. 2004). Towards formal verification of analog designs. In: *Computer Aided Design, 2004. ICCAD-2004. IEEE/ACM International Conference on Computer Aided Design*, pp. 210–217.

[9] Hartong, W., Hedrich, L., and Barke, E. (2002). Model checking algorithms for analog verification. In: *DAC'02: Proc. the 39th Conference on Design Automation*, pp. 542–547. ACM Press, New York.

[10] Henzinger, T. A. (1996). The theory of hybrid automata. In: *Proc. the 11th Annual IEEE Symposium on Logic in Computer Science (LICS 1996)*, pp. 278–292.

[11] Henzinger, T. A. and Raskin, J.-F. (2000). Robust undecidability of timed and hybrid systems. In: Lynch, N. A. and Krogh, B. H., editors, *Hybrid Systems: Computation and Control, Third International Workshop (HSCC 2000)*, vol. 1790 of *LNCS*, pp. 145–159. Springer.

[12] Schönherr, J., Freibothe, M., Straube, B., and Bormann, J. (2004). Semi-formal verification of the quasi-static behavior of mixed-signal circuits by sat-based property checking. In: *1st International Symposium on Leveraging Applications of Formal Methods (ISoLA)*, Paphos, Cyprus.

Chapter 4

IMPROVING EFFICIENCY AND ROBUSTNESS OF ANALOG BEHAVIORAL MODELS

Daniel Platte[1], Shangjing Jing[2], Ralf Sommer[3], and Erich Barke[4]

[1] *Infineon Technologies AG, Munich/Germany*
daniel.platte@infineon.com

[2] *Qimonda AG, Munich/Germany*
shangjing.jing@qimonda.com

[3] *Technical University Ilmenau, Ilmenau/Germany*
ralf.sommer@tu-ilmenau.de

[4] *Gottfried Wilhelm Leibniz University, Hannover/Germany*
eb@ims.uni-hannover.de

Abstract In behavioral simulation, performance and robustness are often crucial issues. This paper presents approaches to improve both simulation efficiency and convergence for bottom-up generated behavioral models of nonlinear analog circuits. The strategy is based on an automated modeling flow and focuses on reformulation as well as restructuring of the underlying equations with respect to sequential equations.

Keywords Behavioral modeling, symbolic analysis, nonlinear analog circuits

1. Introduction

Simulation performance is an important issue in the verification of nonlinear analog circuit designs. A promising approach to reduce high simulation times on system level is bottom-up behavioral modeling. As manual modeling is

S.A. Huss (ed.), Advances in Design and Specification Languages for Embedded Systems – Selected Contributions from FDL'06s, 53–68.

time-consuming and error prone and requires a high level of modeling knowledge, an automated modeling technique is desirable for future verification.

Symbolic analysis [1] offers good opportunities to automatically generate accurate behavioral models from a circuit. The tool Analog Insydes [2, 3] is designed to derive differential-algebraic equations (DAE) directly from a circuit to generate behavioral models based on those equations. Various model reduction techniques [3–5] can be applied to reduce the high complexity of the nonlinear circuit equations. Finally, a behavioral model in an analog hardware description language (AHDL) like VHDL-AMS [6], Verilog-AMS [7], or MAST can be generated from the reduced DAEs.

In the presented research, a modeling flow based on Analog Insydes is used for the model generation. The derived (purely analog) equation-based models are of user-specified accuracy and can be parameterized with circuit parameters. They accurately represent the circuit's behavior also including higher order effects.

Although there are highly efficient model reduction techniques, these behavioral models are of exceptional high complexity. Previous research indicated that the performance mainly suffers from the missing consideration of the used simulation algorithms and the ability of simulators to deal with behavioral models of such high complexity as efficiently as with netlist-based simulations [8, 12].

Nevertheless, device models (e.g. BSIM) show that simulators can deal efficiently with complex DAE. Some "clever tricks" within the device model's implementation prepare the contained equations in an adequate form for numerical analysis. To name just a few of these, there are:

- Procedural evaluation of equations (reducing the number of equations to be solved simultaneously)
- Preevaluation of common subexpressions (to avoid multiple evaluation)
- Convergence aids and limiting functions (to improve robustness and avoid floating exceptions) and
- Approximated derivatives (to reduce complexity)

Some of these strategies may also be applied to behavioral models [9] always assuming the behavioral simulator supports the needed modeling features.

In this paper we present algorithms for the automatic reformulation and restructuring of differential-algebraic equations for modeling purposes. Key issues are the recognition of sequential equations from general DAEs and their usage to improve the simulation efficiency and robustness. Furthermore, unnecessary multiple evaluation of subexpressions during the simulation will be addressed by recognizing common subexpressions and substituting them by additional sequential equations.

The paper is structured as follows: Section 2 describes the used bottom-up modeling flow. A brief introduction to sequential equations will be given in Section 3. In Sections 4 and 5, automated reformulation strategies for DAEs will be presented. The resulting models' efficiency in numerical analyses will be shown in Section 6. Finally, Section 7 summarizes the presented research and points out future aspects.

2. Bottom-Up Behavioral Modeling Using Analog Insydes

Automatic bottom-up generation of behavioral models by symbolic analysis allows to derive accurate behavioral models. The model generation process in this paper is based on the symbolic analysis tool Analog Insydes [2]. This powerful tool offers the functionality to automatically set up circuit equations for a circuit netlist and to use them as basis for a behavioral model. Figure 4.1 visualizes an exemplary process for the bottom-up model generation.

Symbolic device models (corresponding to the simulator's device models) are used to make the strategy as accurate as the circuit simulation itself. The circuit equations are usually set up in an extended modified nodal analysis (MNA) for nonlinear equations. The resulting nonlinear equations contain the network equations in MNA (as used in most circuit simulators) as well as the nonlinear element relations resulting from the device model.

The complexity of the circuit equations is increasing proportionally to the circuit's size, which is again amplified by the complexity of the used device models. Therefore, the achieved equation sets are often extremely complex impossible to set up manually. Model reduction methods can be applied to reduce the complexity of the equations by term reduction techniques [3, 14]. The benefit of this symbolic approximation technique is to ensure a user-specified accuracy. Hence, this is one of the very few methods that allows satisfying a predefined accuracy.

As the equations' complexity decreases while the resulting error increases with the degree of the model reduction, it is up to the user to find a suitable trade-off between size and accuracy of the model. Experiments show that for reasonable error margins (5–10%) the complexity can be reduced by a factor of 10 to 100.

Finally, the behavioral model can be generated from DAEs by using Analog Insydes' model export function. It generates several AHDLs (VHDLAMS, Verilog-A, MAST, etc.) and hence supports the creation of models for almost every behavioral simulator.

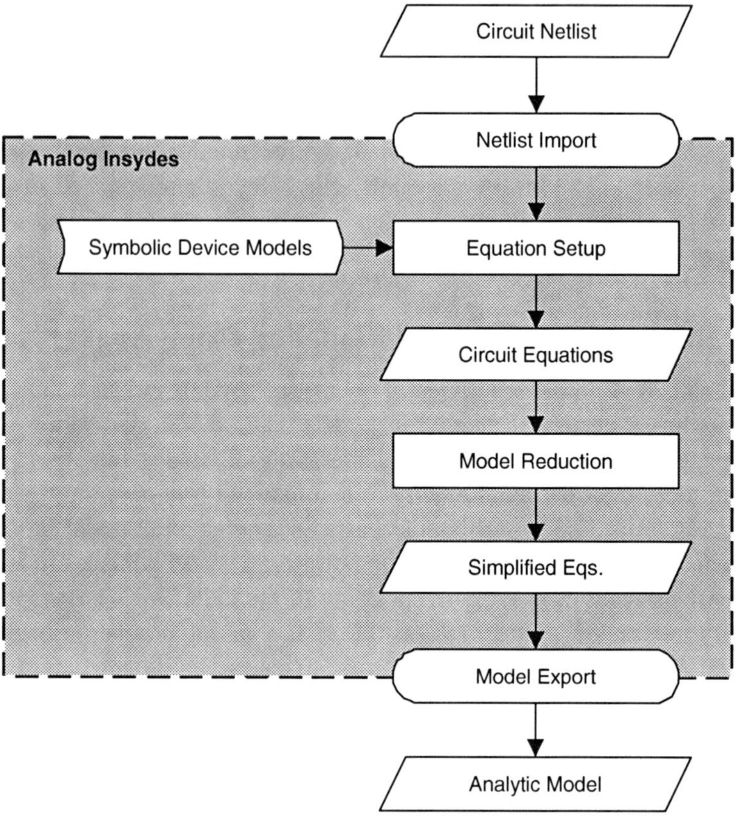

Figure 4.1. Bottom-up modeling process.

3. Modeling with Sequential Equations

From the implementation of device models for common circuit simulators (e.g. SPICE) it is obvious that only a relatively small amount of the contained equations is copied into the simulator's Jacobian matrix and solved iteratively with Newton's method. The main fraction is internally preprocessed in a procedural way and subsequently used to compose the so-called stamp that is needed to set up the system's Jacobian via MNA. The main intention of this handling is to keep the system as small as possible as the complexity for solving a system of linear equations is dominated by its dimension (and sparsity).

The presented reformulation of behavioral equations focuses on handling as many equations as possible in a similar way. We define an ordered set of sequential equations as follows:

Definition 3.1. Sequential Equations $\underline{G}(\underline{x}, \underline{y})$ with
$x_1 = G_1(\underline{y})$ (first equation)
$x_n = G_n(\underline{y}, x_1 ... x_{n-1})$ (any following equation)
where
$\underline{x}$: sequential variables
$\underline{y}$: simultaneous variables

Hence, sequential equations have to be explicitly solvable for their corresponding sequential variable and must only depend on simultaneous as well as previously determined sequential variables. The resulting system $\underline{\sum}(\underline{x}, \underline{y})$ is the union of sequential equations $\underline{G}(\underline{x}, \underline{y})$ and simultaneous equations $\underline{F}(\underline{x}, \underline{y}) = \underline{0}$. The special case of multiple independent blocks of sequential equations could be of advantage for parallel processing, but will not be explicitly treated in this publication.

The (arbitrary) equations Eq. (4.1) to Eq. (4.5) give an example for a system of sequential structure:

$$x_1 = y_1 - y_2 \tag{4.1}$$

$$x_2 = \sqrt{\dot{x}_1 + 1} + x_1 e^{y_1} \tag{4.2}$$

$$x_3 = x_2 - 5x_1 \tag{4.3}$$

$$y_1 + 3y_2 + x_1 e^{y_1} x_3 = 5\sin(t) \tag{4.4}$$

$$\sqrt{y_1} + 2x_2 + x_1 e^{y_1} + y_2 = 0 \tag{4.5}$$

The sequential variables x_1, x_2, x_3 can be determined by evaluating the sequential equations Eq. (4.1) to Eq. (4.3). In fact, they could also be substituted into the simultaneous equations Eq. (4.4) and Eq. (4.5) to reduce the system's dimension. However, substituting all sequential expressions results in expressions of enormous complexity and is of major disadvantage for numerical solving strategies.

Although we do not have any detailed information on how sequential equations are handled in commercial simulators, we suppose that they will not be solved by Newton iteration. The resulting diagonal block in the Jacobian matrix will probably be used by the simulator to reduce the size of the system (e.g. by applying a Schur-like method or eliminating matrix entries). Hence, they will not influence the dimension of the remaining linearized system that has to be solved iteratively. This leads to a reduced complexity for the linear solver. Additionally, procedural evaluation reduces convergence problems during simulation as the values for sequential variables can be accurately determined.

Starting the modeling from a general system of DAE, the information about which equations to handle sequentially and simultaneously respectively has to be introduced. In the next section, this problem will be addressed by a recognition and ordering algorithm. In bottom-up modelling, there is also an alternative way to derive the information about sequential equations: As the equations are set up from a netlist, the used symbolic device models are implemented such that they provide model-internal sequential equations directly.

Once the partitioning into sequential and simultaneous equations/variables is determined, the information has to be passed on to the simulator. In Verilog-A, procedural assignments as defined in Section 6.4 of [7] are suitable for modeling sequential equations. The VHDL-AMS standard also provides a simultaneous procedural statement (see Section 15.4. of [6]), but unfortunately this feature does not seem to be supported by any simulator available to the authors.

4. Recognition of Sequential Equations

Using sequential equations for improvements in modeling was previously presented in [10, 13]. In [11] an algorithm for taking advantage from sequential equations was presented. The recognition was done during the MNA-like set up of the circuit equations in a bottom-up modeling strategy.

In this section, a general algorithm to identify sequential equations from a system of DAEs will be presented. In contrast to [11], this approach is also suitable for manual modeling and more general in the recognition.

The algorithm's objective is to find a partitioning and ordering of the equations of a system of DAEs to be compliant with Definition 3.1 and have as many sequential equations as possible. An additional criterion for a good partitioning is to handle a maximum of the contained nonlinear equations in a procedural way. This helps to prevent convergence problems during the iteration of the remaining simultaneous system. Generating independent blocks of sequential equations is also beneficial but will not be addressed in this context, as it is conflicting with the main objective of finding as many sequential equations as possible.

During the identification process, the equations and variables of the original system are successively separated into sequential and simultaneous sets. Identifying a sequential equation in this context means to

- Identify the equation to be sequential
- Determine its corresponding sequential variable
- Identify and ensure that the equation may be explicitly solved for this variable
- That it can be inserted into the ordered set of sequential equations without conflicting with the desired lower diagonal structure (rf. Definition 3.1)

To ensure that a new sequential equation is explicitly solvable for its sequential variable a symbolic solving method is applied. As several inverse functions are of disadvantage for numerical methods, a user-specified black list of functions not to invert can be provided.

The identification process can be roughly divided into two phases – the identification of independent and dependent sequential equations.

Initially, the algorithm starts with all equations and variables unclassified. Optionally, it is possible to start with a user-defined initial set of simultaneous variables simplifying the identification process and/or an initial set of sequential equations which will be kept during the processing. The algorithm itself is based on the analysis of dependency matrices having the same structure as a Jacobian matrix and indicating for each equation which variables are referenced. To derive the dependency matrix the symbolic Jacobian of the system is set up and filtered for linear, nonlinear, and structural zero entries. Contained derivatives with respect to time can be handled like the variable itself and hence also result in entries within the dependency matrix.

Independent sequential equations do not depend on simultaneous variables. They can be evaluated directly during the numerical solving of the equations. Alternatively, they may be substituted into the remaining equations. A typical example for equations of this type are parameter calculations. They can be identified from the dependency matrix by searching rows containing only one unknown variable – the sequential variable. Once a row satisfying this criterion is found, the corresponding equation is added to the set of sequential equations. Its variable is added to the set of sequential variables and henceforth treated as a known variable. The phase is repeated until no additional equation fulfills the above criterion.

A dependent sequential equation may additionally contain simultaneous variables. As it is advantageous to identify as many nonlinear equations as possible, the remaining unclassified equations are preordered giving the nonlinear equations a higher priority to be used as sequential equations. In this phase, simultaneous variables are also included in the set of known variables. Similar to the previous phase, rows with only one dependency on an unknown variable will be extracted repeatedly. A key feature in this phase is to also identify favorable variables that can be handled simultaneously as this will almost always allow to solve some more unclassified equations sequentially. Every time the identification loop is stuck, the algorithm heuristically determines a variable to be kept simultaneously by the number of unclassified equations depending on the same variable. The higher the number of dependent equations for a variable, the better this variable is suited to be treated simultaneously as this reduces the number of unknowns for as many remaining unclassified equations as possible.

Once all variables have been classified or no more sequential equations can be identified the identification process terminates. All remaining equations and variables are declared simultaneous.

Figure 4.2 illustrates the algorithm in six steps for an arbitrary system (diagram 1) of nine equations (numbered 1 to 9) depending on the variables A to I. Each diagram represents the dependency matrix of the equations on the corresponding variables. Light gray boxes represent linear nonzero entries

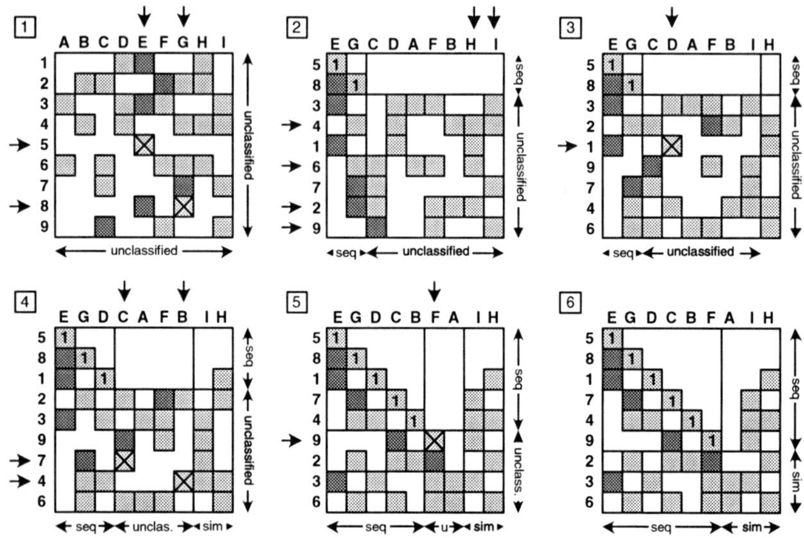

Figure 4.2. Example for the sequential identification algorithm.

of the Jacobian matrix whereas dark gray boxes mark nonlinear dependencies. The white boxes depict structural zero entries. The arrows indicate rows or columns which have been identified to be swapped in the next process step. The used equation and sets of variables are visualized at the right and bottom sides of the matrices. The crosses indicate entries to be identified as sequential variables in the next step. These entries are moved to the correct position by row and column swapping and the corresponding equation is reformulated into an explicit formulation. Hence, the diagonal entry becomes one (assuring good pivot elements).

Within the example, six equations and variables have been identified to be sequentially solvable (see diagram 6). To emphasize some special behavior, two transitions will be discussed in detail:

In diagram 2, the algorithm is stuck as no equation is depending on less than two unclassified variables. To solve this problem, one variable with as many dependencies on unclassified equations is declared to be solved simultaneously. Choosing variable H as simultaneous variable enables the algorithm to identify equation 1 with variable D.

At the stage of diagram 5, both equation 9 or 2 could be used as sequential equations for variable F. As equation 2 contains F in a nonlinear context, it is more useful to select equation 9. Otherwise, the inverse of a nonlinear function would have been needed or (even worse) no explicit formulation would be possible.

The algorithm allows to automatically derive an ordering of general equations and variables, which is of advantage for numerical solving of the system of DAE.

As will be presented in Section 6, the algorithm has been successfully applied to systems of up to 1,600 equations. It can be easily applied in bottom-up modeling as well as in postprocessing of DAEs derived from any other modeling strategy.

5. Identification of Common Subexpressions

Another preprocessing step for generating efficient behavioral models is the extraction of common subexpressions within a system of DAE. The strategy of Common Subexpression Elimination (CSE) is known from compiler design and reduces the computational effort by avoiding the multiple evaluation of common expressions. These expressions are assigned to a temporary variable and referenced multiple times instead of evaluating the expression multiply.

A similar approach can be used for refactoring DAEs: Expressions that are used multiple times in a set of equations do not need to be evaluated repeatedly. After identifying such expressions additional sequential variables will be introduced with the corresponding expressions assigned to the new variables. Consequently, each occurrence of the corresponding expression in the set of equations is substituted by the newly introduced sequential variable.

Within the example given in Section 3, there is a common subexpression that could be represented by an additional y sequential equation $x_4 = x_1 e^{y_1}$. It would be added to the equation system and all occurrences of the subexpression within the original system would be substituted by x_4.

For integrating this strategy into the modeling flow two essential steps had to be performed:

- The recognition of common subexpressions
- Their substitution by additional sequential equations

The first step is based on finding matching subtrees within the expression trees of all equations. The intention is to efficiently find as many maximum matching subtrees as possible.

In an initialization phase, expression trees for the equations are set up. After that, all leaves with multiple appearance of those trees are extracted as a starting set for the algorithm. Hence, this set contains repeatedly used variables, parameters, and constants of the original DAEs.

Recursively, the algorithm then extends the set of subtrees by one level of hierarchy and check, whether the extended subtrees are still contained multiple times. The recursion terminates once all subtrees are processed and no bigger common subexpressions are found.

The second step of the handling for common subexpressions is the substitution. For all common subtrees exceeding a user-specified minimum depth additional sequential equations and variables are introduced and the occurrences are substituted. To keep the sequential structure of the DAEs, each new sequential equation is inserted before its first usage within the equation set. Additional equations which have only been used within the simultaneous equations are inserted at the end of the sequential equation block.

Although the strategy increases the number of (sequential) equations, it reduces the complexity of the original equations and avoids repeated evaluation. Thus, the overall performance during the simulation is increased.

Figure 4.3 shows an exemplary structure of a Jacobian matrix containing nine equations, six of them being sequential. Assuming the algorithm found two common subexpressions in equations 3/5 (being a function of variable B, dark gray) and in 8/9 (function of G/I, black), Fig. 4.4 visualizes the structural changes and insertion points for the new equations 10 and 11. The variables J and K represent the extracted subexpressions and are used to substitute the expressions in 3/5 and 8/9, respectively.

A minor drawback of this strategy may arise if multiple independent blocks of sequential equations should be kept. Identifying common subexpressions used within different blocks and substituting them with an additional equation may

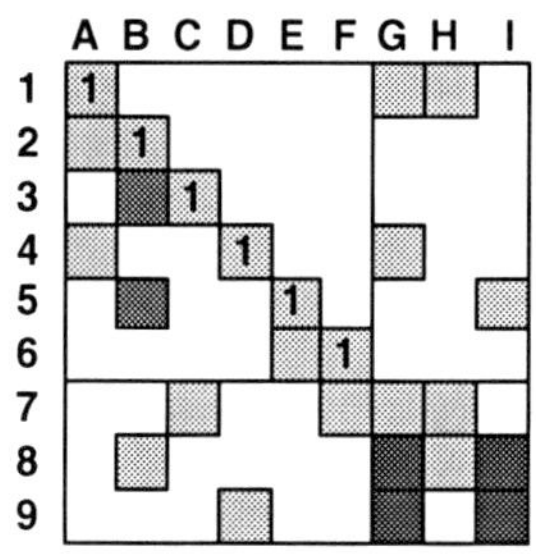

Figure 4.3. Original Jacobian structure.

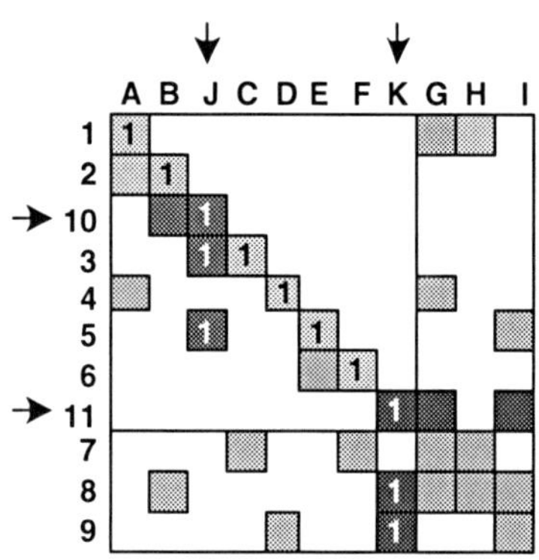

Figure 4.4. Jacobian structure after algorithm.

affect the independence of the blocks in some special cases: If a subexpression is only dependent on parameters and constants, it creates an additional sequential variable that interconnects independent blocks. As partitioning is not (yet) used within our approach, there is no unfavorable effect of such substitutions.

6. Results

In this section, the results of applying the refactoring algorithms described in Sections 4 and 5 will be shown. Front end and back end of the model generation flow used to prepare the equations and behavioral models shown in this section were similar to the flow shown in Fig. 4.1. In contrast to that process, no model reduction was applied. Additionally, the algorithms for recognizing sequential equations as well as the optimization of common subexpressions were applied to the circuit equations. In the modeling process, fully symbolic device models of the dynamic Gummel-Poon bipolar model and the BSIM3v3 MOS model were used to set up the circuit equations. For the presentation of the refactoring results, three exemplary circuits have been selected. Table 4.1 shows some statistics for the examples.

The example opamp741 contains the well-known μA741 operational amplifier. The second example named cfcamp is a complementary folded-cascode operational amplifier designed in Infineon's 130nm technology. Finally, the vco example is implemented as a current-starved ring oscillator (3 stages) and modeled with the full BSIM3 model.

During model generation, the identification algorithm was applied to the circuit equations to identify sequential equations in three different use cases:

(1) Identification starting without initial information (the general case)

(2) Trying to identify additional sequential equations keeping the sequential equations from device models

(3) Additionally constraining the algorithm to keep the block structure of existing sequential equations

Table 4.1. Statistics for the examples.

Examples	Transistors	Device Model	Model Equations	Model Parameters
opamp741	26	Gummel-Poon	368	564
cfcamp	19	BSIM3	1158	1760
vco	16	BSIM3	858	1285

Table 4.2. Identification results for opamp741.

opamp741	Seq. Eqs.	Sim. Eqs.	Reduction
(0) Original Equations	182	186	Reference
(1) Identification without Ini-	252	116	38%
tial Information	(+70)	(−70)	
(2) Identification Keeping	208	160	14%
Original Sequential Eqs.	(+26)	(−26)	
(3) Identification Keeping	206	162	13%
Block Structure	(+24)	(−24)	

Table 4.3. Identification results for cfcamp.

opamp741	Seq. Eqs.	Sim. Eqs.	Reduction
(0) Original Equations	1026	132	Reference
(1) Identification without Ini-	1083	75	43%
tial Information	(+57)	(−57)	
(2) Identification Keeping	1081	77	42%
Original Sequential Eqs.	(+55)	(−55)	
(3) Identification Keeping	1038	120	9%
Block Structure	(+12)	(−12)	

Tables 4.2 and 4.3 summarize the identification results for both circuits. The respective first line shows the ratio of sequential and simultaneous equations resulting from the symbolic device models. In the following rows, the results of the above stated different application modes are shown. The last column presents the reduction of the dimension of the remaining simultaneous equations (relative to (0)).

The results in (1) indicate that even without any initial sequential information the identification achieves the best results. Applying the algorithm to a system of DAE already containing a large part of sequential equations in addition to (1) further reduces the dimension of the remaining simultaneous equations (as shown in (2)). A minor drawback is that the independency of the blocks (originally: opamp741 – 26×7 seq. eqs.; cfcamp – 19×54 seq. eqs.) cannot be retained. Constraining the algorithm to retain the block structure (see (3)) reduces the efficiency of the identification algorithm.

The recognition rate for the vco example is given in Table 4.4 and confirms the efficiency of the algorithm.

The elimination of common subexpressions can preferably be used after the identification of sequential equations. Table 4.5 shows that for both examples a relatively high number of repeatedly used expressions was extracted and defined in additional sequential equations.

Finally, the effect of this refactoring on the simulation performance was measured. The generated Verilog-A models were simulated using Cadence's AMS Designer (Spectre kernel) and the CPU time of the transient analysis was used for the benchmarking of the refactoring results.

For all examples no convergence was achieved when modeling all equations simultaneously. Possible reasons for this are the extremely high number of equations and the (especially for the BSIM3 model) highly nonlinear contained

Table 4.4. Identification results for vco.

opamp741	Seq. Eqs.	Sim. Eqs.	Reduction
(0) Original Equations	756	102	Reference
(1) Identification without	787	71	30%
Initial Information	(+31)	(−31)	
(2) Identification Keeping	796	62	39%
Original Sequential Eqs.	(+40)	(−40)	
(3) Identification Keeping	766	92	10%
Block Structure	(+10)	(−10)	

Table 4.5. Results of common subexpression elimination.

	Seq. Eqs.	Sim. Eqs.	Total
	206	162	368
opamp741	787	71	30%
	(+31)	(−31)	
	1038	120	1158
cfcamp	1753	120	1873
	(+715)		(+715)
	796	62	858
vco	766	92	
	(+10)	(−10)	10%

Table 4.6. Simulation performance after refactoring.

Performance	Original Equations	Identification	Identification and CSE
opamp741	9.07 s	8.06 s	7.17 s
	Reference	(−11 %)	(−20.9 %)
cfcamp	3.9 s	3.5 s	1.5 s
	Reference	(−10%)	(−61.5%)
vco	760 s	504 s	233 s
	Reference	(−34%)	(−69%)

expressions. When sequential equations were used in the model generation, good convergence, and identical simulation results to the corresponding netlist-based simulation have been obtained. Table 4.6 summarizes the performance improvements for both examples. The identification (3) of additional sequential equations resulted in a speedup of 11% for the opamp741. Extracting 104 common subexpressions with the presented algorithm further improves this to a total speedup of 21% compared to the simulation containing only the sequential equations derived from the symbolic device models (0).

For the cfcamp, even better results have been achieved: The identifaction of sequential equations resulted in a reduction of 12 simultaneous equations (9%) and improved the simulation efficiency by 10%. Extracting common subexpressions additionally introduced 715 new sequential equations. In combination with the sequential identification algorithm, the CPU time for the transient simulation was significantly reduced by 61.5%. The simulation performance of the vco example was enhanced by 69% by preprocessing the equations. These statistics show that the speedup caused by identifying sequential equations is roughly proportional to the reduction rate of the linear system's dimension. Improvements through common subexpression elimination are scaling somehow proportionally to the number of saved expression evaluations.

7. Conclusions and Future Aspects

In this publication, two algorithms for preprocessing large systems of DAEs for efficient and robust modeling have been presented: The identification of sequential equations from DAEs allows to use procedural statements for modeling the major portion of typical bottom-up generated circuit equations. Hence, the number of equations to be solved with iterative methods is reduced significantly resulting in increased efficiency and robustness in numerical analysis. The algorithm presented in Section 5 extracts common subexpressions within DAEs and introduces additional sequential equations to avoid multiple numerical evaluation of those subexpressions during simulation. Both methods were integrated into an automated bottom-up modeling flow based on symbolic analysis. The efficient application for refactoring DAEs has been demonstrated considering as example circuit blocks of typical size. The achieved results show that the methodology can successfully be applied in model generation and significantly improves the simulation performance. Future aspects of this research cover the automatic insertion of convergence aids into model equations as well as partitioning sequential equations into independent blocks for further performance improvements. Another aspect is the realization of a sequential solving strategy for Infineon's inhouse simulator Titan to support the usage of VHDL-AMS simultaneous procedural statements.

References

[1] G. Gielen and W. M. C. Sansen. *Symbolic Analysis for Automated Design of Analog Integrated Circuits*, Kluwer Academic Publishers, Boston, MA, 1991

[2] Analog Insydes: www.analog-insydes.de

[3] T. Wichmann and M. Thole. Computer Aided Generation of Analytic Models for Nonlinear Function Blocks, 10th Intern. Workshop PATMOS, September 2000

[4] C. Borchers. Symbolic behavioral model generation of nonlinear analog circuits, *IEEE Trans. on Circuits and Systems II: Analog and Digital Signal Processing*, 45(10), October 1998

[5] C. Gathercole and H.A. Mantooth. Pole-zero localization: A behavioral modeling approach, *Proc. the Fifth IEEE Intern. Workshop on Behavioral Modeling and Simulation*, October 2001

[6] IEEE Standard 1076.1, "IEEE Standard VHDL Analog and Mixed-Signal Extensions", March 1999

[7] Accellera, "Verilog-AMS Language Reference Manual", Version 2.2, November 2004

[8] D. Platte, R. Sommer, and E. Barke. An approach to analyze and improve the simulation efficiency of complex behavioral models, *Proc. the IEEE Intern. Behavioral Modeling and Simulation Conference*, 2006

[9] G. J. Coram., How to (and how not to) write a compact model in Verilog-A, *Proc. the IEEE Intern. Behavioral Modeling and Simulation Conference*, 2004

[10] L. Näthke, V. Burkhay, L. Hedrich, and E. Barke. Hierarchical automatic behavioral model generation of nonlinear analog circuits based on nonlinear symbolic techniques, *Proc. the IEEE Design, Automation and Test in Europe Conference and Exhibition*, 1(1), 2004

[11] L. Näthke, L. Hedrich, and E. Barke. Betrachtungen zur Simulationsgeschwindigkeit von Verhaltensmodellen nichtlinearer integrierter Analogschaltungen, 6. ITG/GMMDiskussionssitzung Entwicklung von Analogschaltungen mit CAE-Methoden, May 2002

[12] D. Platte, S. Jing, R. Sommer, and E. Barke. Ansätze zur Verbesserung der Simulationsperformance automatisch generierter analoger Verhaltensmodelle, 9. ITG/GMM/GI Workshop Methoden und Beschreibungssprachen zur Modellierung und Verifikation von Schaltungen und Systemen, February 2006

[13] L. Näthke. Ansätze zur automatischen Generierung hierarchischer Verhaltensmodelle von nichtlinearen integrierten Analogschaltungen, Logos Verlag, Berlin, 2004

[14] T. Wichmann. Symbolische Reduktionsverfahren für nichtlineare DAE-Systeme, Shaker Verlag, Aachen, 2004

Chapter 5

MODELLIB: A WEB-BASED PLATFORM FOR COLLECTING BEHAVIOURAL MODELS AND SUPPORTING THE DESIGN OF AMS SYSTEMS

Torsten Mähne and Alain Vachoux
Laboratoire de Systèmes Microélectroniques (LSM)
Ecole Polytechnique Fédérale de Lausanne (EPFL)
Bâtiment ELD, Station 11
CH-1015 Lausanne
Switzerland
torsten.maehne@epfl.ch
alain.vachoux@epfl.ch

Abstract This chapter describes ModelLib, a web-based platform for collecting models from different domains (e.g. electrical, mechanical, or electromechanical) and levels of abstractions (e.g. system, circuit, or device level). Use cases for this tool are presented, which show how it can support the design process of complex heterogeneous AMS systems through better reuse of existing models for tasks like architecture exploration, system validation, and creation of more and more elaborated models of the system. The current state of the implemented ModelLib prototype is described and an outlook on its further development is given.

Keywords Heterogeneous SoC, AMS design process, model library, model meta-information, relational database, Apache, PHP, subversion, wiki.

1. Introduction

Systems-on-Chips (SoCs), as combinations of computer and communication hardware and software equipped with autonomy based on perception, cognition, and control capabilities, are key parts of a perpetually broadening range of

S.A. Huss (ed.), Advances in Design and Specification Languages for Embedded Systems – Selected Contributions from FDL'06, 69–88.

applications, from industrial equipment to personal appliances. The design of SoCs has currently to address a number of significant issues:

Increasing complexity, due to the integration of significant computing and communication power (intelligent systems).

Significant heterogeneity, due to the variety of integrated components (analogue/RF/digital hardware, embedded software, sensors, actuators).

Increasing environmental awareness, due to energy saving, battery-operated systems, environmental monitoring capabilities, and continuous interaction with the working environment.

Increasing impact of modern silicon technologies, due to deep sub-micron and nanometer technological processes.

Increasing reuse of subsystems, due to ever shrinking time to market and rapid product obsolescence.

The fast progressing advances in manufacturing technology allow the integration of more and more functionality from different disciplines into a single complex heterogeneous SoC. This leads to a continuous growth in the needed design effort, where at the same time product cycles get shorter. The resulting increase in the "design productivity gap" is especially notable in semiconductor industry. There, the technological production capacity (measured by the number of available transistors) has increased since 1985 yearly between 41% and 59%, whereas the design capacity (measured by the efficient use of transistors) has increased only at a yearly rate of 20% to 25% [9]. To allow the control of the design costs and to prevent them to get prohibitively expensive, new design technologies have to be continuously introduced, like block reuse or IC implementation tools.

The design of heterogeneous Analogue and Mixed-Signal (AMS) systems is still a highly manual work and not as automatised as the design of digital systems using logic synthesisers and place and route tools. Their design requires a diversity of description formalisms, also called Models of Computation (MoCs), analysis and simulation methods provided by specialised simulators for different physical disciplines and levels of abstraction, as well as CAD tools for the design and layout of the physical realisation of each heterogeneous system component. For example, the design of a typical Micro-Electro-Mechanical System (MEMS) like an inertial yaw rate sensor [8] requires, among others:

- Optimisation and characterisation of the micromechanical resonator and (separately) of the electrostatic field distribution of the comb drive structures, which are driving and sensing the movement of the flexible structure, using a FEM tool like ANSYS from mechanical engineering

- Simulation of the whole system on the circuit level, taking into account the coupling between mechanical and electrostatic domain within the MEMS transducer and feedback from the analogue and digital driving and sensing circuits using behavioural simulators and modelling languages like VHDL-AMS from electrical engineering
- Layout of the mechanical structure and of the electronic circuits using IC layout tools

These tools originate from different engineering fields, which leads to problems when exchanging models and other design data. The designers are forced to bridge the gaps between tools and methodologies using manual conversion of models, proprietary tool couplings, and tool integrations. It is still difficult to handle all the different design aspects of a mixed-signal system simultaneously. An efficient tool support for the AMS design flow (Fig. 5.1) is thus missing and rendering it overly complicated, error prone, and time consuming.

The challenge for the EDA industry in the short term is to improve the links between the existing tools. Here, one research area, which relates to the given yaw rate sensor example, is the development of reduced-order modelling methods that allow the extraction of fully coupled behavioural models for circuit level simulation from detailed FE models of the device [8]. In the long term, new design methods and integrated tool chains are needed to support the whole process of specification, design, integration, validation, verification, and integration of the components of a complex AMS system. First approaches for the specification, synthesis, and automated layout generation exist for moderate-complexity analogue circuits (device count less than 100), e.g. the *AMGIE* approach and the *Mondriaan* tool described in [16].

Behavioural modelling of analogue/RF/MEMS blocks is a methodology that is being increasingly used in industry today during the design of integrated systems. Behavioural models describe the functionality of a component as input–output behaviour augmented with major non-idealities of real implementations,

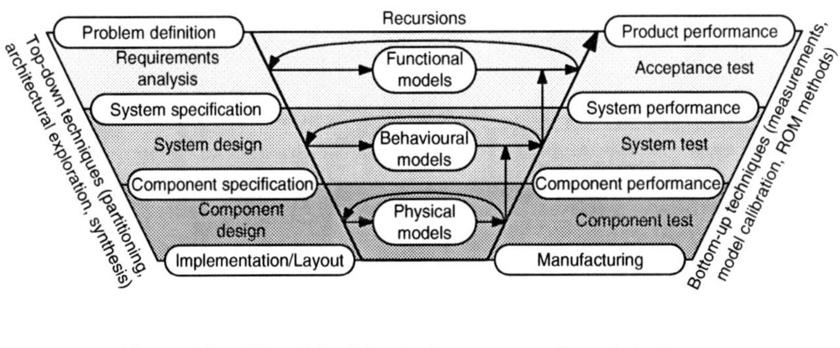

Figure 5.1. V-model of the design process of a technical system.

but without requiring a complete description of real implementations. The purpose is mostly to verify the correct functionality of the system in acceptable CPU times by replacing (some) analogue/RF circuit schematics or 2D/3D models with behavioural models. The usage of behavioural models is beneficial on both sides of the V-shaped design process (Fig. 5.1):

- During *top-down design*, to develop a system architecture that meets the system specifications. Although automatic formal refinement or synthesis methods for analogue/RF/MEMS systems are still mostly lacking, the existence of a *well-defined library of behavioural models* and *area/power estimation models* combined with fast high-level simulation methods may considerably help the designers during architecture exploration and allow optimal mapping of the top-level performance specifications among the different blocks in the system architecture, trading off different specifications and implementation cost (e.g. area, power). An example is given in Fig. 5.2, which shows a possible realisation of the receiver front end of a digital telecommunication link and the block specification parameters, which have to be derived from the overall system specifications.

- During *bottom-up design*, to allow the overall system verification, where the detailed circuit netlists are replaced by more abstract models that are characterised against their respective circuit implementations. The behavioural models for that kind of task may be the same as the ones used in the top-down phase or they may be different, if they are obtained by extracting/simplifying the circuit or device behaviour using the already mentioned reduced-order modelling methods, which involve typically symbolic equation manipulation or the generation of look-up tables.

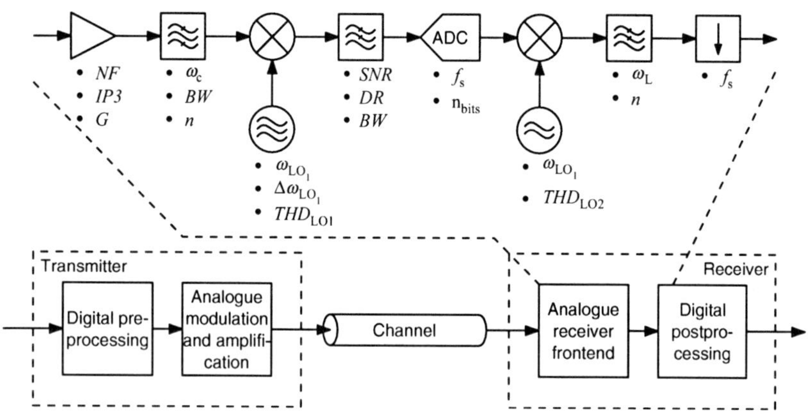

Figure 5.2. Choosing receiver front end architecture and deriving block specifications of a digital communication link (adapted from [3]).

The management of models of a device, component, or the whole system on different abstraction levels is thus an important aspect of an AMS design flow.

The behavioural modelling is facilitated through standardised mixed-signal and mixed-technology behavioural description languages like VHDL-AMS or Verilog-AMS [13], which are implemented by several commercial simulators. The SystemC-AMS modelling framework [15] extends the capabilities of VHDL-AMS at the system level by supporting Models of Computation (MoCs) that allow the development of abstract models and efficient simulation of complete heterogeneous systems. However, the development of a model for a specific block is not an easy task because of several issues:

- The *block heterogeneity* requires specific knowledge of all the different involved physical domains.
- The model is the *formalisation of the designer's intent* and is thus requiring from him the knowledge of the modelling language, the methodology, and the used design tools.
- The model needs to be *adapted to the tools and for the specific design task*, e.g. simulation, (formal) verification, optimisation, or synthesis.
- The *abstraction level needs to be adapted to the specific design task* to choose the right trade-off between level of detail (accuracy) and the execution speed of the model.

To speed-up the creation of these models, it is advantageous to reuse existing models and to possibly adapt them further. Nowadays, designers often reuse only their own models or those provided by the design environment for two important reasons, which need to be addressed. First, an exchange of the models between designers is complicated by the fact that they are often not aware if and where a similar model is already existing. Second, the designer has to gain trust in the validity of the model for his specific design task, which is more difficult to achieve for foreign models because of the "Not invented here" syndrome. To overcome these problems, the models need to be documented regarding their interface, implementation, extend of covered effects, how they were verified, and other general properties. The designer needs to understand from this information the model structure and its functionality as well as to judge, if it is suitable for a given task. It has to be clear with which tools the model is expected to be compatible. The Engineering Society for Advancing Mobility Land Sea Air and Space International (SAE) covers the documentation issues in the *SAE J2546 Model Specification Process Standard* [14]. There are ongoing activities to develop collections of verified models for different design languages like *Modelica* [6] and *VHDL-AMS* [4]. Those libraries are often available as archives from the Internet. They usually provide some documentation besides the model source code, which can be, to some degree, automatically extracted from the model sources using a documentation generator.

There are also tool-specific library managers, e.g., the *Library Manager* in the Cadence IC Design Environment or the Workspace in Mentor Graphics ModelSim, to handle the various models of a project and the design-kits. However, these tools cannot cope with the management of the models over the tool boundaries, as it is required for heterogeneous AMS system designs. Handwritten websites documenting and linking to model archives are a solution, but can only partly address these issues and require a lot of manual maintenance to stay up to date. The ModelLib project addresses these problems through the development of a web-based platform with the following objectives:

- Creating a platform for collecting and distributing models independently of the design tools
- Establishing a validation process for the collected models through collaborative review and development
- Supporting the designer's decision for the right model for each task (e.g. architecture exploration, performance analysis, verification)
- Realising this in an open-source framework, but with appropriate IP protection for the stored documentation and models

This chapter describes the ModelLib platform that is being developed to address the described needs. It will also offer features supporting the design of complex heterogeneous AMS systems. Section 2 presents the basic use cases for a model library and how it can support the work of the AMS designer. From this, requirements for the ModelLib platform are developed. The architecture of a prototype implementation is described in Section 3. Conclusions are given together with an outlook on further developments in Section 4.

2. Use Cases and Requirements for a Model Library

A model library like ModelLib can be set up on different organisational levels, like within a project group, a company, or as a community portal on the Internet. The basic use cases (Fig. 5.3) for the AMS designer accessing the ModelLib server through a client on his computer for submitting, retrieving, and collaboratively developing the models over the Internet remain the same, while the demands for security and required detail of access control will rise with each level of broader access. The communication between client and server needs to be done through an encrypted channel. Users have to authenticate themselves in front of the model library so that the information stored in the library can be selectively made available to the different users. This is needed to protect the Intellectual Property (IP) of the authors and to support the conformance to their

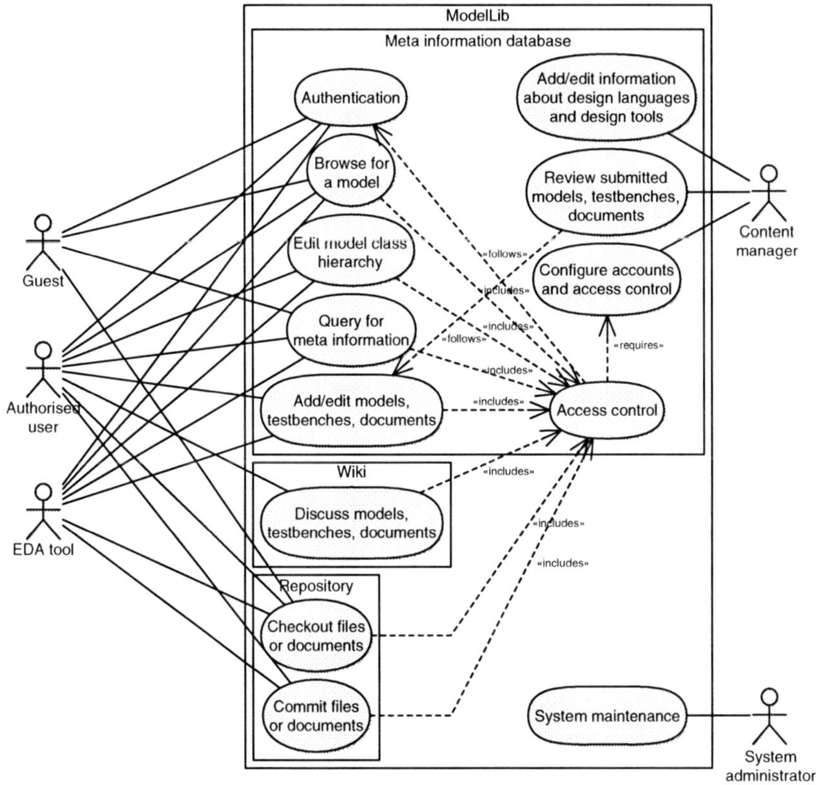

Figure 5.3. Use cases of a model library.

licence terms, under which they are making their work available to the public or a restricted group of users.

Users of the model library can be categorised into five different roles, each with a different profile of allowed actions on the library (Fig. 5.3). The *guest* is anonymous and has thus the minimum rights. He can only browse for a limited collection of public models, send queries about their meta-information, and retrieve their source code by checking it out from a public repository. By authenticating himself with a valid user name and password, he can become an *authorised user*, who gains further rights dependent on his membership in different groups. These groups *grant* or *deny* him access to the different parts of the library (the different models, testbenches, and documents themselves, as well as the supplementary meta-information about them). He can participate in the review process by discussing the models, testbenches, and documents stored in the library. He can also actively contribute to their development by submitting new models, adding/editing of the meta-information about them, and organising them into different categories called *model classes*. Also *EDA tools* need also a direct access to the model library and are a special form of an authorised user. There are two privileged roles. The *content manager* configures accounts and access rights of the users, reviews their submitted models, testbenches, and documents, and adds commonly used information about supported design languages and tools. The *system administrator* is responsible for the maintenance and the further development of the model library platform.

In the following, the use cases for retrieving a model (through browsing or a more complex query) are discussed in more detail, since they show which meta-information need, to be stored in the library alongside the models to support the designer's decisions. Then follows the description of the remaining use cases regarding submission, updating, and jointly developing models.

One way to access the models in the library is to directly browse through the collection of available models. For this they need to be sorted into a hierarchy of *model classes*. A model can be at the same time member of different classes, e.g. to reflect that a model is part of some IP library and that it is modelling effects from a particular physical domain. In [14, Appendix A] a list of automotive EE commodities structured into a hierarchy of EE Commodities, Class, Sub_Class I, and Sub_Class II is given. It illustrates the diversity of automotive EE commodities that may be subject to modelling and simulation. It could serve as a guide to structure the model library. After selecting a model, the user is presented the meta-information describing the properties of the model, which can be detailed into *structural meta-information* describing "How the model is built?" and *semantical meta-information* describing "How the model can be used?". The *structural meta-information* describes the following aspects:

- *Name and storage place* of the model within the model class hierarchy
- *Interface*, including detailed information about all parameters and ports as well as the assertions associated to it
- One or more *model implementations (architectures)* in the form of behavioural description(s) and/or structural description(s) along with the assertions associated to it/them
- *Design entities*, each one gathering the interface and one model implementation using a particular design language (e.g. VHDL-AMS) and tested against particular tools (e.g. simulators or synthesisers)
- *Testbenches*, which are stored with the models and refer to some design entities. They are implemented using a design language version and stored in a number of files, which are known to work together with some design tool versions. Test data and expected results can be included.

To each of these aspects, an arbitrary number of references to external documents can be given. These are information, which can be directly extracted from the model sources. To support the porting of design entities and testbenches to other design tools, it is required to store additionally information about the different versions of available design languages and design tools, as well as which tool versions support which language versions. Figures 5.6, 5.7, and 5.8 from the ModelLib prototype described in Section 3 show one way of presenting the structural meta-information to the user.

The *semantical meta-information* further characterises a model regarding its fidelity, performance, and usage. The *SAE J2546 Model Specification Process Standard* [14] discusses some of these aspects. It recognises that models can be classified according to their sophistication, capability, and captured intelligence and that the many dimensions of fidelity make the classification challenging, since these dimensions may be sequential, parallel, independent, contradictory, and/or redundant. Properties of a model may be related to the entire *model*, a particular *feature* it implements, or the way of its *execution*. The semantical meta-information includes for example:

Model refinement level: The SAE J2546 standard proposes a number of model refinement levels, of which some are of interest in the context of a model library:

Pins: The model consists only of an interface with ports, parameters, and assertions. An instance of the model can be created, but cannot be executed since the model does not implement any internal feature.

Static: The model implements time-invariant, steady-state internal behaviour using some primary quantitative properties suitable for DC or steady-state AC analysis. An amplifier could be represented on this level through a controlled source; a piezoresistive pressure sensor only through its stiffness, the stress-free resistance, and the piezoresistive coefficient.

Dynamic: The model implements time-varying behaviour, possibly including non-linear characteristics, but neglecting smaller second order effects to capture only the primary time-dependent behaviour. For example, the amplifier model could include its limited bandwidth and a slew rate. A pressure sensor model could include, stress-stiffening, inertia, and damping effects.

Precision: The model implements a significant amount of second order effects in addition to the effects necessary to capture its primary time-varying behaviour. For an amplifier model, this could mean that it includes noise and thermal effects like self-heating and temperature-dependent gain variation. A pressure sensor model could include cross-sensitivity to other physical quantities like temperature as well as fatigue or hysteresis effects.

Vector: The model implements directional or spatial interfaces to its environment in contrast to the previously one-dimensional lumped connection points. This could be achieved, e.g. through a distribution function or multiple connection points. Typical examples are models with spatially distributed parameters, e.g. a MOSFET device model taking into account its geometry and doping profiles or a 3D FE model of a pressure sensor.

Model of Computation (MoC): on which the model is based, may it be a discrete MoC (Discrete-Event (DE), Finite State Machine (FSM), Petri net, Data Flow (DF), ...), continuous MoC (signal flow, conservative network, bond graph, Finite Element Method (FEM), ...), or in some way a synchronised mix of discrete and continuous MoCs.

Physical disciplines: the function of a modelled component is based on, electrical, mechanical, thermic, hydraulic, optic, etc.

Validity of the model: describes under which assumptions and operating conditions a model is valid.

Suitability for design tasks: e.g. architecture exploration, area/power estimation, connectivity verification, or bottom-up verification (using back-annotated or calibrated models).

Keywords: to index the model.

Feature properties: A model has individual *features*, each capturing a different aspect of the component with a varying level of detail: the *nominal behaviour* (e.g. amplification or signal filtering) of the component, plus *secondary behaviour* (e.g. temperature dependency or noise) caused by its interaction with the operation environment. A captured model feature can be mapped, e.g. through global variables, the parameters, or the ports of the instantiated model on the internal variables of a model.

A feature is often based on a *physical effect*, e.g. variation with T, manufacturing tolerances, aging, self-inductance, or noise. It influences the *performance criteria* of a component, e.g. filter characteristic, filter order, cut-off frequency, bandwidth, sample frequency, or the noise figure.

The *feature refinement level* describes how far a model feature is implemented (levels 0–7 in SAE J2546): *None* (not included), *Named* (acknowledged but unimplemented), *Fixed* (adjustment only through editing the model or a non-related parameter), *Index* (offers discrete choice of discrete values or modes), *Static* (accepts parameter value prior simulation run), *Dynamic* (adapting to internal conditions during simulation), *Mutual* (adapting to external influences during simulation), *Directional* (adapting to directional external influences during simulation).

Execution capabilities: Depending on its implementation, a model is suitable for different types of execution, notably *simulation* and *synthesis*. The results obtained from the model execution need to be characterised.

A model can support for *simulation* different *analysis types* like *Continuity* and *Loads Analysis*, *Nominal Analyses* (DC, transient, small signal, and stability analysis), *Stress Analysis*, *Perturbation* or *Sensitivity Analyses*, *Worst-Case Analyses*, *Failure Modes and Effects Analyses (FMEA)*, or *Sneak Circuit Analysis (SCA)*. For each analysis type, the model can give different results (e.g. current, voltage, temperature, force, power, failure), obtainable either dynamically during or only after completion. Each result can be in a different form like *Flag*, *Message*, *Scalar*, *Waveform*, or *Relation* (non-time series, describing in the form of, e.g. an equation or a table of the interaction of two or more variables).

Synthesis transforms a model through an algorithm into another model. During top-down design, details are added in each synthesis step. For example, a program implementing an *algorithm* (C, MATLAB, VHDL, etc.) can be transformed into an Register Transfer Level (RTL) description (VHDL, Verilog, etc.), then into a *gate level* description (SPICE or Verilog netlist), and finally into a layout. During bottom-up design, models can be simplified through reduced-order modelling methods to make them suitable for simulation on higher levels of abstraction.

It is possible to include all this supplementary information into free-form description fields, but for large model collections, like they are intended for ModelLib, it is better to structure them as far as possible and to store them in an adapted data structure. This has to be done as general as possible because not all properties, which a designer might think of storing for a model, are known in advance. It allows the AMS designer to better manage the models for different levels of abstraction of the same physical component.

A large collection of models also requires a more efficient access to the models, besides browsing, to support the designer in selecting the model with the right fidelity for reuse in the design tasks (e.g. architecture exploration, detailed component characterisation, system validation). To do that, the designer has to send queries to the model library. In simple cases, this means querying for a *model name*, *key words*, and full text search in the description fields of the interfaces, architectures, etc. If this is not sufficient enough, more complex queries for certain properties of the model to be in a specified range (e.g. detail level, modelled effects, design language, component properties) can be made.

The use cases for browsing and querying of models showed which information has to be provided by the model developer when submitting a new model to the library. First, the files containing the source code, testbenches, test data, simulation results, and other documents of the model must be made accessible through Uniform Resource Locators (URLs), preferably by submitting them to the repository of a revision control system. This eases their further (cooperative) development, keeps track of the modification history of each file, and keeps thus all model revisions accessible. Then, the meta-information about the model, testbenches, etc., needs to be extracted from these files and entered into the library using structured input forms. This process can be partially automatised using similar techniques like the ones used by documentation generation tools [4]. During the further development of the model, the meta-information needs to be continuously updated to keep it in sync with the changes made. The library supports the development process by providing a forum for discussing the model and jointly improving its implementation and documentation.

The newly submitted model is stored, in the beginning, in one of the private model classes of the model developer, for which he can specify the access rights. To publish the model to the other users by including it into the official collection of the library, he has to contact one of the content managers. They will do a first review to evaluate if the model has reached a level of quality to be made public. This review includes checks if the model source code is following some suggested coding rules (e.g. no syntax errors, well-structured and well commented code, consistent naming scheme, ...) as well as if the supplied meta-information and documentation is correct (consistent with the model source code) and complete enough (permitting understanding and usage of the model by a third party). The content manager gives feedback to the author until the requirements for the publication of the model are met. Then, the new model is sorted into the appropriate public model classes and announced to be available under certain licence conditions. This formalised submission process shall insure a basic quality standard of the available models in the library. Subsequently, other authorised users can comment on the different parts of the model, can give a rating of the model, and can contribute to its further development by providing patches and documentation. Rating charts of the models may be implemented to motivate the developers through a sense of competition. The goal of these measures is to create an active community of model developers, which help and motivate each other while constantly improving the model library.

3. Implementation of the ModelLib Prototype

Several software components have to be integrated to meet the requirements arising from the use cases presented in Section 2. The core component of ModelLib is a database that stores the meta-information about the models. All files that contain the models and their accompanying documents are managed by a revision control system to allow their collaborative development.

A wiki provides an open platform to discuss the models and collaborate on the improvement of their implementation and documentation.

A running prototype of ModelLib [7] is being implemented, which is using *PostgreSQL* [12] to manage the database that stores the meta-information, *Subversion/WebSVN* [2] to handle the model repository, *YaWiki* [5] to discuss and jointly develop the documentation of the models, and *PHP* [11] to implement the web interface served out by an Apache 2 web server [1]. This section describes the current state of the prototype.

Figure 5.4 shows the architecture of the prototype and how its different components interact. The lower part of the figure shows the different user-created documents managed by the ModelLib server. The file revisions of the documents are stored in the *repository*. The meta-information about models and accompanying documents are stored in the *meta-information database*. Informal texts, like discussions and HOWTOs, are stored in the *wiki database*. On the server side implemented user interfaces provide the access to these three storages, which has the advantage that users can access ModelLib over the Internet using a standard web browser. The file revisions of the models and documents are managed on the user side by the Subversion client. It also provides the possibility to commit them to and update them from the repository.

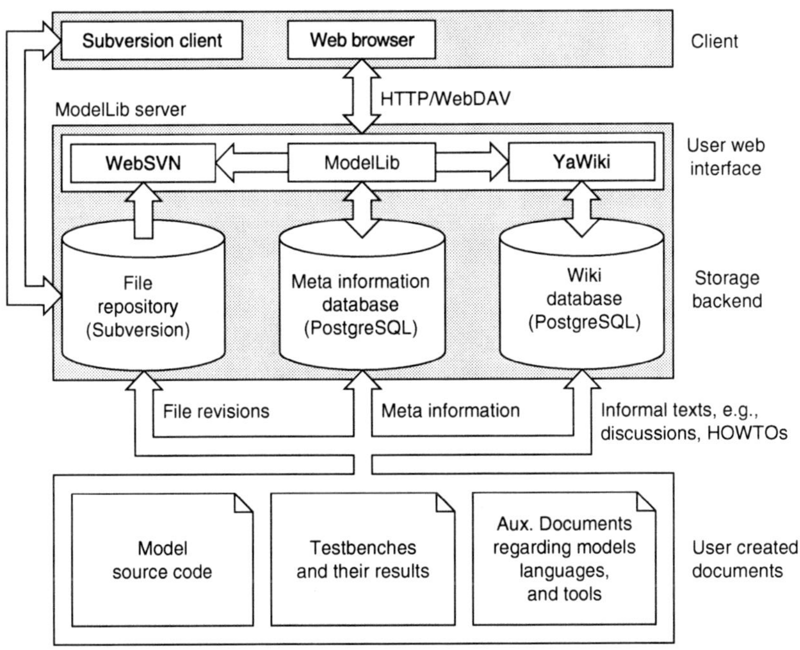

Figure 5.4. Architecture of the ModelLib prototype and the communication links between its different components.

The meta-information describing the properties of the models is stored in a relational database. Figure 5.5 shows the Entity-Relationship (ER) diagram of the database that has been designed for the prototype. It currently considers the model class hierarchy, the information about referenced external documents, the information about available design language and tool versions, and the meta-information about interface, architectures, assertions, design entities, testbenches, and files of the models. This fully covers the structural meta-information described in Section 2. The structural meta-information is not yet implemented in the database structure and can be stored, currently, only as free-form descriptions. PostgreSQL [12] has been chosen to implement the database for the ModelLib project because of its (among others) full ACID compliance, good ISO SQL standard coverage, and its handling of foreign key constraints.

The ModelLib web interface provides access to the model collection over the Internet. It queries the meta-information about models, testbenches, design languages, design tools, and document references from the Relational Database Management System (RDBMS) and outputs it to a CSS formatted HTML page. In the prototype, it is implemented in the server side scripting language PHP, using library packages from the public PEAR repository [10], which ease the development of web applications. For example, the ModelLib prototype uses `DB` and `Text_Wiki` to access the database on the PostgreSQL server and format the HTML output of the free-form description fields using a wiki-like syntax. Currently, the web interface implements the following features:

- Logging in as different database users
- Browsing the model class hierarchy for a model
- Displaying and editing of all information related to the interface, architectures, design entities, and testbenches of a model (Fig. 5.6)
- Displaying of the information about the available design languages and their different versions (Fig. 5.7)
- Displaying of the information about the available design tools and their different versions (Fig. 5.8) and
- Displaying/editing of the document references

The files containing the models, testbenches, and their accompanying documents are stored in a repository, which is managed by a revision control system to allow their collaborative development. The meta-information database refers to the files in the repository using URLs. The *Subversion* system has been chosen for the ModelLib prototype since it provides most features of the widely

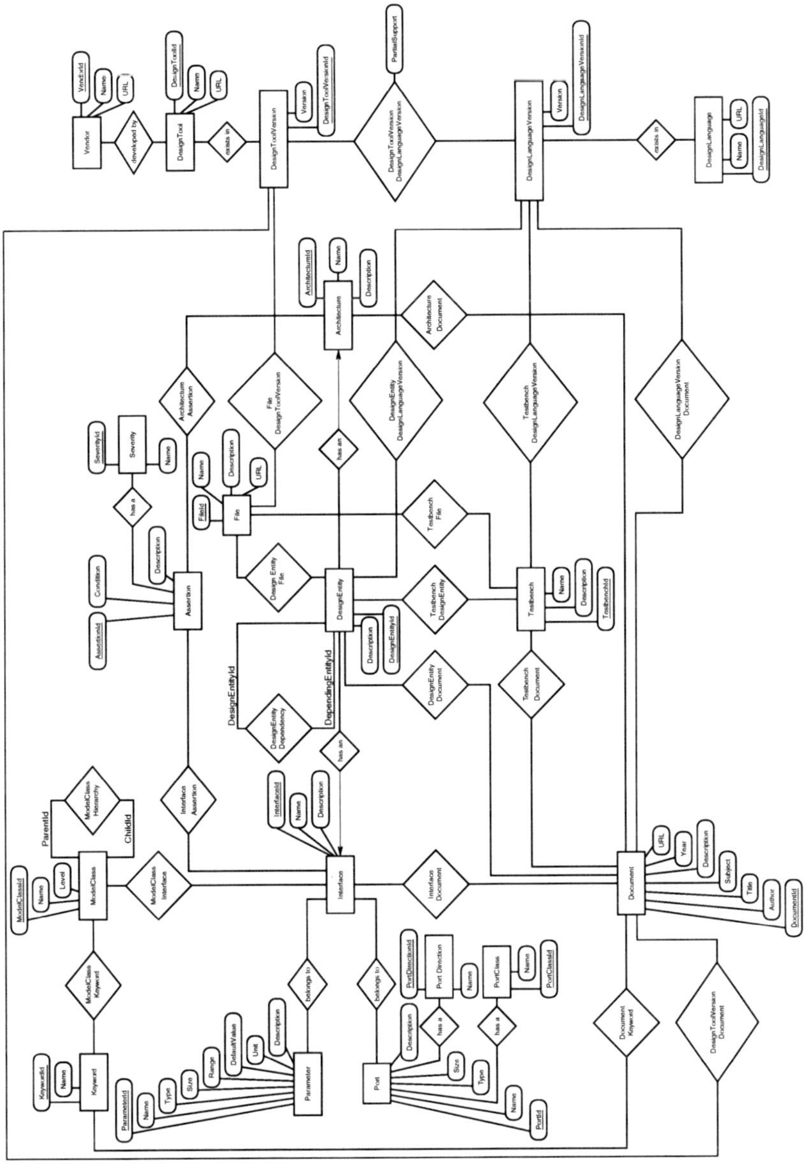

Figure 5.5. Entity-Relationship diagram of the meta-information database.

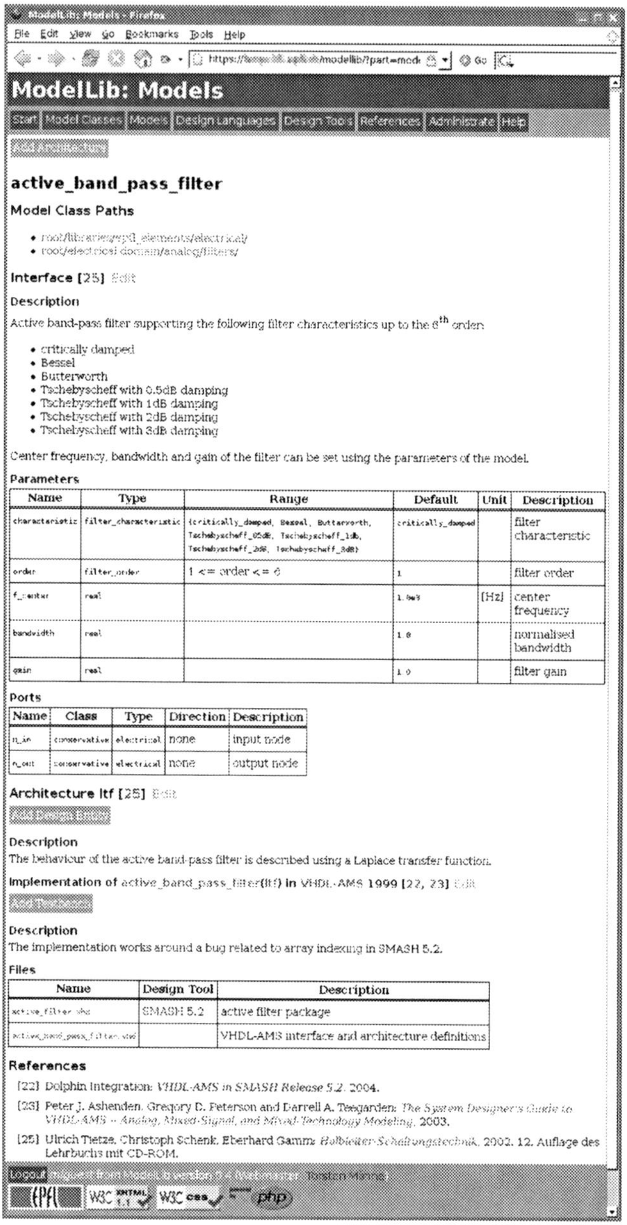

Figure 5.6. Meta-information about a selected model.

used CVS and overcomes some of its known drawbacks by supporting, among others, versioned directories, renames, and meta-data; truly atomic commits; efficient handling of binary files; and more efficient handling of tags and branches. An Apache 2 web server, using the module `mod_dav_svn.so` from

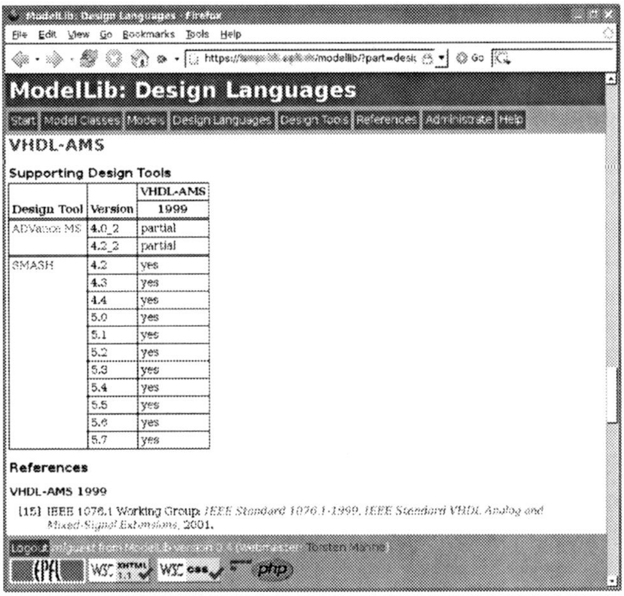

Figure 5.7. Meta-information about design languages.

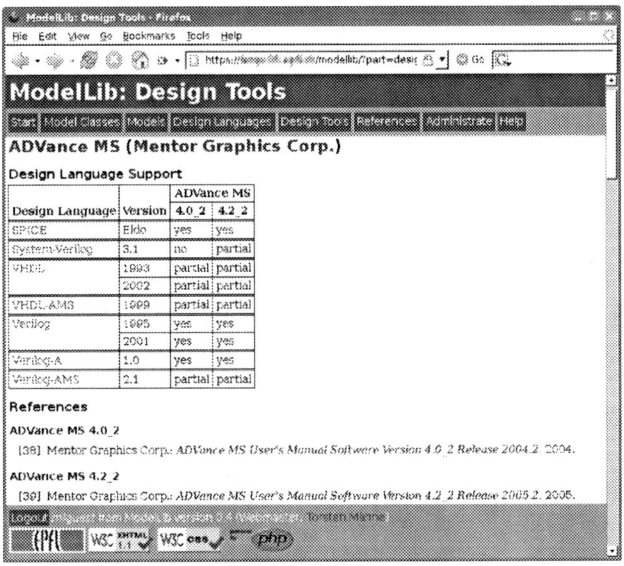

Figure 5.8. Meta-information about design tools.

Subversion, makes the repository available to the clients via the WebDAV protocol [17], which is an extension to the HTTP 1.1 protocol that adds versioned-writing capabilities. This provides key features like authentication, path-based authorisation, wire compression, and basic repository browsing using a web browser or WebDAV client. The PHP-based web interface *WebSVN* improves the browsing of the repository via a web browser considerably, with features like viewing the file/directory logs; listing of all changed, added, or deleted files in a queried revision; log message searching; Blame support; Tar ball downloads; Directory comparisons; and RSS Feed support.

A wiki provides an open platform to discuss the models and to collaborate on their implementation and documentation. An access rights management has to be established to control the read and write access to the different models. *YaWiki* [5] was chosen for the ModelLib prototype because it adds logical name spaces, Access Control Lists (ACLs), navigational elements, and more to the traditional wiki; each wiki page can be instantly commented through a web form; it is written in PHP like the other parts of the ModelLib web interface; its formatting engine `Text_Wiki` is a PEAR module, which is also used in the ModelLib web interface to format the free-form description fields.

4. Conclusions and Outlook

The ModelLib prototype presented in Section 3 implements basic features of a model library as described in Section 2. Its web interface allows browsing for a model through the model class hierarchy. The meta-information about models, testbenches, design languages, design tools, and external documents that are stored in the meta-information database can be displayed. New models can be added by committing them into the repository and adding/editing their meta-information in the meta-information database using the web interface.

Currently under development is a unified authentication and a fine-grained access control mechanism using ACLs to allow the public usage of ModelLib over the Internet, while keeping control of who has read and write access to the different parts of the library. The web interface is being reimplemented on the base of the Java Platform, Enterprise Edition (Java EE), to improve the modularity of the data storage, application logic, and presentation layers and to realise direct EDA tool access through an Application Programming Interface (API). This will also give the opportunity to improve the usability and to complete the functionality of the web interface. The three main components, i.e., repository, meta-information database, and wiki, will be better integrated to offer the user a coherent interface. The main task for the further development of the ModelLib prototype is to support the structured storage of the semantical meta-information, which characterise the fidelity, usage, performance, and other properties of the model. This implies an extension of the current database scheme. The querying of models is currently only supported through direct SQL queries to the meta-information database. This use case needs to be implemented into the web interface, so that it will offer the designer elaborated query schemes to guide him/her in the selection of a suitable model for his current task. The efficiency of the model import and update can be improved by automatising the input of the meta-information in the ModelLib database by extracting the information from the model source code.

In the near future, the approach presented in this chapter shall be validated through the support of practical design cases such as the design of RF transceivers and multichannel A/D converters.

References

[1] The Apache Software Foundation. *The Apache HTTP Server Project*, 1999–2005. http://httpd.apache.org/, visited on 27th March 2006.

[2] B. Collins-Sussman, B. W. Fitzpatrick, and C. M. Pilato. *Version Control with Subversion*. O'Reilly Media, Inc., Sebastopol, USA, 1st edition, June 2004. http://svnbook.red-bean.com/, visited on 27th March 2006.

[3] G.G.E. Gielen. CAD tools for embedded analogue circuits in mixed-signal integrated systems on chip. *IEE Proceedings – Computers and Digital Techniques*, 152(3):317–332, 2005.

[4] E. Hessel, K. Panreck, J. Haase, A. Schneider, and S. Scholz. Development of VHDL-AMS-libraries for automotive applications. In: *Forum on Specification and Design Languages (FDL) 2005*, pp. 101–110. ECSI, September 2005.

[5] P. M. Jones. *YaWiki: Yet Another Wiki for PHP*, 2005. http://www.yawiki.com/, visited on 27th March 2006.

[6] The Modelica Association. *Modelica: Modeling of Complex Physical Systems*, 1997–2006. http://www.modelica.org/, visited on 30th March 2006.

[7] T. Mähne. *ModelLib Prototype*. Laboratoire de Systèmes Microélectroniques (LSM), Ecole Polytechnique Fédérale de Lausanne (EPFL), July 2006. https://lsmpc4.epfl.ch/modellib/, visited on 13th July 2006.

[8] T. Mähne, K. Kehr, A. Franke, J. Hauer, and B. Schmidt. Creating virtual prototypes of complex MEMS transducers using reduced-order modelling methods and VHDL-AMS. In: *Applications of Specification and Design Languages for SoCs: Selected papers from FDL'05*, The ChDL series, pp. 135–153. Springer, 2006.

[9] I. O'Connor, F. Tissafi-Drissi, G. Révy, and F. Gaffiot. UML/XML-based approach to hierarchical AMS synthesis. In: *Forum on Specification and Design Languages (FDL) 2005*, pp. 89–100. ECSI, September 2005.

[10] The PHP Group. *PEAR: The PHP Extension and Application Repository*, 2001–2005. http://pear.php.net/, visited on 27th March 2006.

[11] The PHP Group. *PHP: Hypertext Preprocessor*, 2001–2006. http://www.php.net/, visited on 27th March 2006.

[12] PostgreSQL Global Development Group. *PostgreSQL: The World's Most Advanced Open Source Database*, 1996–2006. http://www.postgresql.org/, visited on 27th March 2006.

[13] F. Pécheux, C. Lallement, and A. Vachoux. VHDL-AMS and Verilog-AMS as alternative hardware description languages for efficient modeling of multidiscipline systems. *IEEE Transactions on Computer-Aided Design of Integrated Circuits and Systems*, 24:204–225, 2005.

[14] SAE Electronic Design Automation Standards Committee. SAE J2546: Model specification process standard. Technical report, SAE: The Engineering Society For Advancing Mobility Land Sea Air and Space International, 400 Commonwealth Drive, Warrendale, PA. 2002.

[15] A. Vachoux, C. Grimm, and K. Einwich. Extending SystemC to support mixed discrete-continuous system modeling and simulation. In: *Proc. the IEEE International Symposium on Circuits and Systems (ISCAS) 2005*, pp. 5166–5169, 23–26 May 2005.

[16] Geert Van der Plas, Georges G., and Willy S. *A Computer-Aided Design and Synthesis Environment for Analog Integrated Circuits*. Kluwer Academic Publishers, Dordrecht, The Netherlands, 2002.

[17] J. Whitehead. *WebDAV Resources*, 2005. http://www.webdav.org/, visited on 27th March 2006.

II

C/C++-BASED SYSTEM DESIGN

Introduction

The five papers selected for this part of the book are not only excellent contributions to the C/C++-based System Design (CSD) workshop, but also reflect the different aspects covered in the workshop. These aspects include modelling and design techniques for simulation, debugging, transformation, and analysis of hardware/software systems. The lion share of the papers use C/C++-based languages like SystemC.

The first paper presents an approach called "Quiny" which allows to analyse and transform SystemC models by introspection at run-time. The approach is illustrated by a SystemC to VHDL translation. In the second paper the methodology and a tool are described to extract metadata from (a library of) SystemC IP components. The metadata describe the (hierarchical) structure and communication interface of components to enable system level IP-composition. The techniques presented in the third paper enable more efficient debugging of SystemC designs without changing the models. With minor modifications to the SystemC kernel the GNU debugger has been extended by high level debugging commands. The case study presented in the fourth paper demonstrates how TLM techniques can be applied in HW/SW co-design to shorten the design time. The experiment shows that the required changes in design methodology are acceptable compared to the benefits. The final paper in this part describes a mechanism to combine object-orientation with OSCI TLM by introducing "active objects". A small case study illustrates the use and results of the approach.

I hope that this brief introduction will draw your attention to the scientific articles in the following chapters and stimulate you to participate in the next CSD workshop at FDL 2007 in Nice.

Frank Oppenheimer
OFFIS e.V.
R&D division Embedded Hardware/Software Systems

Chapter 6

THE QUINY SYSTEMC™ FRONT END: SELF-SYNTHESISING DESIGNS*

Thorsten Schubert[1] and Wolfgang Nebel[2]

[1] *OFFIS e.V.*
thorsten.schubert@offis.de

[2] *Carl von Ossietzky University Oldenburg*
wolfgang.nebel@informatik.uni-oldenburg.de

Abstract In this article we address the problem of parsing SystemC designs. We present how a reflection-like technique – as an alternative to classical parsers – can be used to obtain a complete representation of a SystemC design. Such a representation can be the basis for further analysis, transformation or synthesis tasks. We illustrate this technique by means of an automatic SystemC to VHDL translation. Furthermore, we compare our approach to classical parsers.

Keywords SystemC parser, reflection, introspection

1. Motivation

SystemC emerged as promising approach for system-level design coping with the ever-increasing complexity of today's and tomorrow's systems. Technically SystemC is a C++ class library which provides a discrete event simulator, specific data types, and an infrastructure for describing hierarchical designs and communication on different levels of abstraction. This makes SystemC an attractive platform for research on new design methodologies. For example, new (experimental) language constructs can be introduced by simply extending the library. Since there is no need for a separate simulator, the extensions

*This work was done in the ICODES project and supported by the European Commission.

S.A. Huss (ed.), Advances in Design and Specification Languages for Embedded Systems – Selected Contributions from FDL'06, 93–109.

can be experimented immediately by just compiling and executing the design under test in conjunction with the extended library. When it comes to automating certain tasks of the design process, however, the problem of processing, i.e. reading and transforming SystemC designs arises. Especially, parsing SystemC is challenging and many approaches have been proposed to tackle this problem.

The rest of this article is structured as follows: Section 2 describes and analyses the problem of parsing SystemC and Section 3 takes a closer look at existing approaches. In Section 4 we present a new technique to (partially) solve the problem and discuss its pros and cons in Section 5. We conclude with Section 6.

2. Problem Statement

In order to build tools which can analyse, transform, synthesise, or perform any other kind of processing of SystemC designs it is obviously necessary to read the SystemC source code as input and to build an appropriate data structure for further processing.

Goals

Depending on the concrete analysis or transformation task, a subset of the following properties of a SystemC design have to be determined:

- The hierarchical structure of a SystemC design: (sub-)modules, ports signals, processes
- A process' sensitivity, accessed ports, statements, and expressions
- The types of ports, signals, variables, etc.
- Type information about user-defined types, i.e. data members, methods, constructors, destructors, and inheritance relations

As an exemplary *application* to illustrate our approach we chose the task of transforming a given RTL SystemC design into an equivalent RTL VHDL description. That is, the input is SystemC source code and the output is VHDL source code.

In order to realise such a translator we need access to all of the properties listed above except for the last one. This can be achieved with an intermediate representation (IR) as shown in Fig. 6.1. The IR is a representation of a whole SystemC design after the elaboration phase as opposed to an abstract syntax tree (AST) which only reflects the static source code structure.

Further potential application scenarios are, for example, documentation generation, visualising a design's structure, state-machine visualisation, fixed-point to integer conversion, or behavioural synthesis.

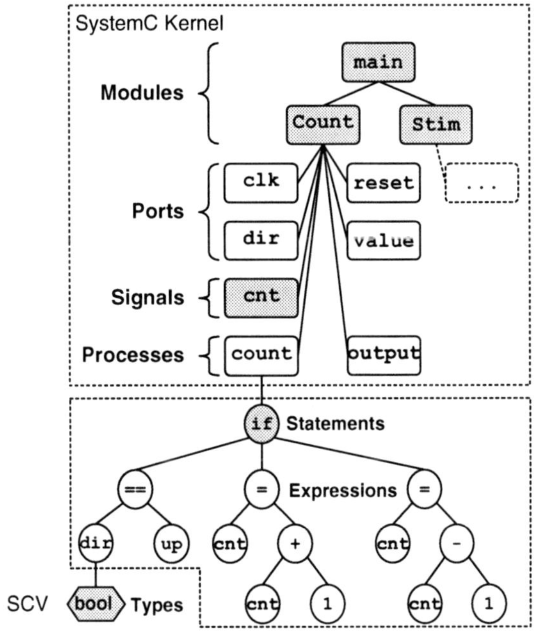

Figure 6.1. Intermediate representation of a SystemC design.

Analysis

The task of creating an AST from source code is usually done by a parser. The creation of parsers is well understood and even largely automated. However, parsing SystemC is still a complex and challenging task. The reason for this is that SystemC is not a language on its own, but a C++ class library. SystemC models which are solely intended to be used for simulation can in fact use all C++ features. As correctly stated in [11]

> Any tool using a dedicated grammar for SystemC would have to include all the grammar and typing rules of the C++ standard in the tool to have a correct parser.

What makes the situation even more complicated is the fact that an AST does not provide all necessary information. It is usually an abstract representation of the static source code structure. In order to infer the hierarchical structure of the design, it is necessary to mimic the elaboration phase, i.e. to perform some kind of interpretation of the code located in the constructors.

Even restricting to hardware synthesis, i.e. imposing certain constraints on the input concerning the allowed language features and code structure, does not significantly lower the complexity. This is due to the fact that (a) a tool provider usually wants to provide a powerful and convenient to use synthesis subset and (b) in principle nearly every C++ feature except for those constructs involving dynamic (de-)allocation of memory, unrestricted usage of pointers, and exception handling can be synthesised. An example for such an extended synthesis subset especially regarding the object oriented features can be found in [5].

The specific problem we address in this article is the creation of the IR from the SystemC source code. The VHDL generation is pretty straightforward since input and output are RT-level models and hence no (complex) transformations are required. However, using a more complex backend – which is not the focus of this article – our approach enables truly self-synthesising SystemC designs.

For our purposes we assume the following requirements on a tool (also called front-end) which creates such an IR:

- Large accepted language subset (ideally arbitrary SystemC models)
- Non-intrusiveness: no parser-specific constructs have to be inserted into the user's code
- Modularity: user-defined backends, convenient API for accessing the IR, extensibility to new language constructs
- Generation of meaningful error messages for ill-formed programs
- Efficiency
- Low costs, i.e. implementation effort and/or royalties

3. State of the Art

In principle, two different categories of approaches can be distinguished: (1) those parsing SystemC source directly and (2) those making use of the run-time features of the SystemC kernel. The former category can further be distinguished into those approaches treating SystemC as *language* having certain keywords and the ones treating SystemC designs as C++ programs adhering to a certain style and using predefined classes.

Parsers

Commercial SystemC tools, which need detailed information about SystemC designs, most probably rely on commercial C++ frontends like [2]. This can also be an option for research projects depending on the constraints on licensing conditions and the available development resources.

In [1] a doxygen-based SystemC parser is described. The XML output available from doxygen is processed and the structure of the design is inferred.

The SystemPerl tool-suite [12] includes a lex/bison/Perl-based parser component which provides basic information about the SystemC design, but does not handle process bodies.

The SystemC2Verilog translator [7] also uses a lex/yacc-based approach to parse SystemC source code. Due to its focus on translating RTL SystemC to RTL Verilog, rather than being a SystemC parser, its subset is limited. For example, module instantiations have to be done via `new`, signal, and port accesses have to use the `read` and `write` methods, and no C data types are allowed.

The SystemC parser ParSyC [3] is also written from scratch using the Purdue Compiler Construction Tool Set (PCCTS/ANTLR). The parser is able to determine the structure as well as the behaviour defined by the method bodies. The supported language subset is not explicitly stated in [3].

KaSCPar [4] is a SystemC parser which was written from scratch using JavaCC and JJTree. Its output is an AST and an XML representation of the SystemC design after elaboration.

Reflexive Approaches

As stated in Section 5.2.20 of the SystemC language reference manual [9], the method `sc_module::get_child_objects()` can be used to obtain a vector of:

> [...] every instance of class `sc_object` that lies within the module in the object hierarchy. This shall include pointers to all module, port, primitive channel, unspawned process, and spawned process instances within the module and any other application-defined objects derived from class `sc_object` within the module.

Note, that this also includes hierarchical channels as these have to be derived from the class `sc_module`. Hence, it is possible to extract the structure of a given SystemC design by simply compiling and running the model until the end of the elaboration phase and then querying its structure. However, neither does this method allow us to analyse the processes on statement- and expression-level, nor does it provide information about non-`sc_object` objects like local variables or data members along with their types.

Regarding the possibility to examine types at run-time, the SystemC Verification Library [13] offers interesting capabilities, i.e. data introspection. Data introspection basically allows a program to deal with types as run-time objects. For instance, a program can query a variable's type, check whether it is a `struct` and then query its elements and further query their types, names, bit-widths, etc. This technique allows for writing very generic code and was proposed for writing transaction-level testbenches. Therefore, the SCV does not immediately help for the given parsing problem, but gives an idea on how to extend the reflective capabilities of SystemC.

Hybrid Approaches

The Pinapa approach [11] can be considered as hybrid approach. It uses the SystemC simulation kernel in order to determine the structure of the SystemC design after elaboration and a modified gcc in order to parse the process bodies at statement- and expression-level. In a final step, both models are combined into a complete intermediate representation of the whole SystemC design.

Discussion

The parser approaches listed in Section 3.0 in principle are able to meet all the requirements listed in Section 2.0 except the last one: A parser which is able to determine all the properties listed in Section 2.0 requires considerably high implementation effort, which is caused by the complexity of C++ underlying SystemC.

The reflexive approaches cited in Section 3.0 are quite limited with respect to the properties they are capable to derive from a SystemC design. However, they do not have to cope with the complexity of C++ the way parsers do. Instead, reflexive approaches even benefit from the expressiveness of C++: The features of a design which shall be extracted later, e.g. its structure, are directly integrated into the modelling library and hence, *make use* of C++.

The Pinapa approach combines the advantages of both approaches. The implementation effort is reduced by a high degree of reuse, thereby making itself dependent on the reused components.

We extend the reflexive approaches concerning the extractable design features while keeping their advantage of dealing with the complexity of C++ with moderate implementation effort.

Referring again to Fig. 6.1 we will focus on the middle part of the IR, that is the extraction of the statements and expressions.

4. The QUINY Approach

The approach we propose in this article is a completely reflexive or, more precisely, run-time approach. That is, QUINY is a library which is linked against the user's code instead of a stand-alone tool. For our exemplary SystemC to VHDL translation tool built on top of QUINY, the usage is illustrated in Fig. 6.2. The user takes the SystemC design, compiles, and links it using the QUINY library instead of the original SystemC library. When starting the resulting executable, the original design prints itself out as equivalent VHDL code.

The execution of the user model in conjunction with the QUINY library and a backend, i.e. the execution of `synth.x` shown in Fig. 6.2, basically consists of two phases: a build-up phase which roughly corresponds to the

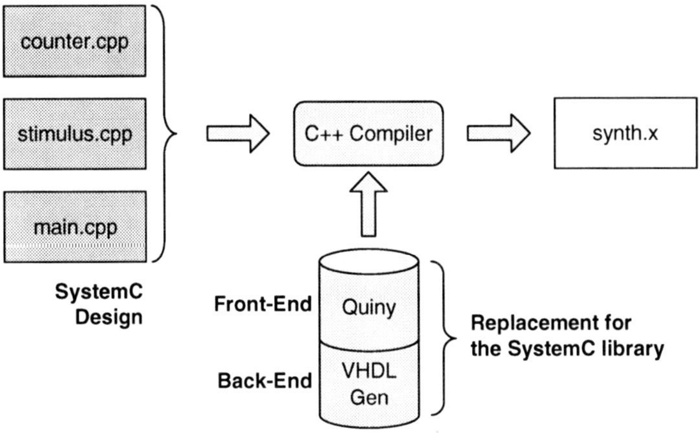

Figure 6.2. QUINY usage.

elaboration phase of the original SystemC simulation and a code generation phase performed by the backend.

The phrase "quiny SystemC front end", or QUINY for short, is a conjunction of "Quine" and "tiny" SystemC front end. A Quine is a program which prints out its own source code. More thoughts on such self-referential systems can be found in [6] by which QUINY was inspired.

Another way to look at it is to say replacing the original SystemC library by the QUINY library can also be seen as replacing the standard simulation semantics by translation or synthesis semantics.

Note that there are two kinds of QUINY users. Firstly, there is the end-user which, for example, wants to do a SystemC to VHDL translation of a certain design. Ideally, the fact that the translation "tool" was implemented using QUINY should be totally transparent to him or her.

The second class of QUINY users are the backend developers. They use QUINY as SystemC front end to obtain an IR of a given SystemC design. The IR is then the basis for the custom analysis, transformation, or synthesis task.

Also note, that QUINY completely replaces the SystemC library, i.e. the header files and the link library. The complete replacement is especially necessary for the scenario where the end-user can take an unmodified design for use with a QUINY-based tool. This scenario is called unintrusive mode.

Unintrusive Mode

As stated in Section 3.0 SystemC already provides an infrastructure for querying a design's structure. Hence, the remaining problem is to extract the structure of the processes on statement and expression level for which C++ does not offer any built-in features. Looking from the C++ point of view at a SystemC design, an SC_METHOD is a member function of a struct or class, which is derived from sc_module. The body of such a method operates on user-defined (as seen by the compiler) template classes like sc_in<T>, sc_signal<T>, sc_uint<N>, and so on.

Expressions. The central idea behind QUINY is to redefine the SystemC classes in such a way that, for example, a signal read expression like m_Count.read() does not produce a value, but a special expression object (EO). Therefore QUINY provides its own definition of sc_signal<T> including the read() method[1]:

```
template<typename T>
Expression&                // The return type is not T,
sc_signal<T>::read()       // but Expression&
{
  // Return an IR node
  return *(new SignalReadValue(this));
}
```

Additionally, EOs provide constructors which allow the compiler to do implicit type conversions. For instance, simple integer literals like 42 or even explicitly converted ones like sc_uint<8>(1) are automatically converted to EOs which represent the corresponding constant expression. Furthermore, EOs have properly overloaded operators. For example, the operator + which takes two EOs as parameters and yields a further EO as result. The resulting EO then represents the sum of the two expressions – as opposed to actually evaluating its value. Since C++ allows to overload nearly all operators, in this way, EOs can be created for arbitrarily complex expressions. The construction and destruction of these EOs can be monitored by QUINY and used to build up an internal expression tree which reflects the original code.

Statements. In order to obtain a process' statements some more work is necessary. The first thing to find out is which of the methods are actually processes. Fortunately, SystemC requires the user to explicitly mark these methods via macros. QUINY uses these macros as hook and defines own versions for them. Basically, the QUINY versions of the macros each create a process node in the IR. This node is registered as current process, which is necessary for QUINY-internal bookkeeping. After that, the corresponding method is called.

During the execution of the method it is necessary to detect statements like if and while instead of actually executing them with their original meaning. This is achieved by macros which redefine these keywords. The very first approach was to create special statement objects similar to the expression objects described in the previous section. For example, such an if statement object would have a constructor that expects an expression object which represents the condition and would look like this:

```
#define if(cond) {IfStatement *tmp=new IfStatement(cond);}
```

Hence, when the compiler processes the code, there are effectively no control structures. The process is simply executed once – creating and destroying various expression and statement objects.

The main drawback of this approach is that it is unable to correctly detect the statements which make up the following then part and the optional else part. This is due to the fact that the following statement can be a single statement or a compound statement enclosed in curly braces {}.[2] Unfortunately, it is not possible to redefine the curly braces via macros or some other mechanism in order to detect the beginning and the end of a block. What makes the situation even more complicated for if-statements is the fact, that the else-part is optional. Especially in nested if-statements it is tricky to associate an else part to the correct if-statement.

One possible workaround is the (mis-)use of a for statement. We exploit the property that a variable or an object can be declared in the initialisation part and that its lifetime matches the execution of the loop body, which in turn can be a single statement or a compound statement. If we can ensure that the loop body is executed exactly once, we do have the desired effect. Hence, a better if-replacement looks like this:

```
#define if(cond)  for(BlockHelper b(BLOCK_IF, (cond)); \
                      b.runOnce == false; \
                      b.runOnce = true)
```

The effect of this macro and a similar replacement for else is shown in Fig. 6.3.

In addition to the control structures, we need to detect the declaration of local variables. Such declarations of SystemC types like sc_uint<N> can be

```
if(a<b)
{
  x = 0;
  y = 0;
}
else x = 100;
C++;
```

```
for (BlockHelper b(BLOCK_IF, (a<b));
     b.runOnce == false;
     b.runOnce = true)
{
  x = 0;
  y = 0;
}
for (BlockHelper b(BLOCK_ELSE);
     b.runOnce == false;
     b.runOnce = true) x = 100;
C++;
```

Figure 6.3. Effect of the macros contained in the QUINY header files: before (left) and after (right) standard C++ preprocessing.

detected by providing QUINY-specific replacements of these classes. Instances of these classes can either be local variables or temporary objects. In order to distinguish local variables like `sc_uint<8> myVar(1);` from temporary objects which are created in expressions like `x+sc_uint<8>(1)`, it is necessary to carefully monitor an object's usage. If the very same object is referenced more than once, it must be a named variable. Unfortunately, the names of local variables cannot be recovered[3] – as opposed to ports, signals, and modules, which carry a run-time name.

To keep track of variables of built-in types like `int`, QUINY provides replacement classes and defines macros which map the built-in types to their replacements. Unfortunately, the built-in types can consist of multiple tokens like `unsigned int` which cannot be handled by our macro approach in unintrusive mode. Furthermore, we currently see no unintrusive way to correctly identify array and pointer declarations. Examples for such declarations are:

```
sc_int<8>            myArray[3];
sc_int<8>           *myPointer;
unsigned long int    myULong;
```

Although it is possible to overload the `operator *` and `operator []` in user-defined classes that operate on the instantiated *objects*, we see no way to influence the *declarator* `*` or `[]`. Only in the special case of heap-allocated arrays, however, a possibility would be to use the `operator new []` as hook.

Intrusive Mode

The unintrusive mode has the advantage that it does not require any intervention by the end-user beyond exchanging the SystemC library in the compile command. However, the unintrusive mode has some limitations which can only be overcome with a little help from the end-user. First of all, it is necessary to include a special header file in the end-user's design. Furthermore, the QUINY data types have to be used instead of the native ones. For example, the macro `Q_ULONG` has to be used instead of writing `unsigned int`. This macro serves as hook – the same way as the macro `int` did in the unintrusive mode. Furthermore, no C/C++ keywords are redefined by macros in unintrusive mode: The end-user must use macros like `IF` and `WHILE` instead of `if` and `while`. Hence, it is possible to escape the "quinyfication" of certain parts of the code, e.g. the constructor code of a module. In the normal SystemC simulation mode these macros are simply expanded to their original meanings.

Another issue which can be solved by the intrusive mode, is the detection of pointer and array variables. By requiring the user to use special array and pointer classes which are provided by QUINY and whose instances exactly behave like the built-in arrays and pointers. For instance, the declarations from the previous example would look like this when using the intrusive mode:

```
Array<sc_int<8>, 3> myArray;
Pointer<sc_int<8> > myPointer;
Q_ULONG             myULong;
```

Implementation

As can be seen in Figs. 6.4 and 6.2, an executable of a SystemC design compiled and linked against QUINY consists of three parts: the end user's SystemC design, the QUINY library, and a backend. To enhance the modularity, the QUINY library and the backend library are two independent link libraries. Hence, for our exemplary SystemC to VHDL translation, the user compiles the design with QUINY's SystemC headers and links it to the QUINY library as well as to the VHDL code generation backend.

The actual control flow is also shown in Fig. 6.4. Execution starts in the `main` function, which is located in the QUINY library. After internal initialisation, it calls the function `sc_main` which is part of the user's SystemC design. Executing the `sc_main` function triggers the elaboration phase in which QUINY builds up its internal data structures through the mechanisms described in the previous sections. Calls to `sc_start` are also intercepted by QUINY. When `sc_main` is finished, QUINY passes control to the backend by calling `quiny_start`.

The most important parameter passed to the backend is `root`. It is the entry point to the IR. An excerpt of the class hierarchy of the IR nodes is shown in Fig. 6.5. The simulation commands `sc` are a kind of container for all `sc_trace` and `sc_start` calls. In our exemplary VHDL generation backend, these commands are converted to a ModelSim script file with equivalent logging and start commands.

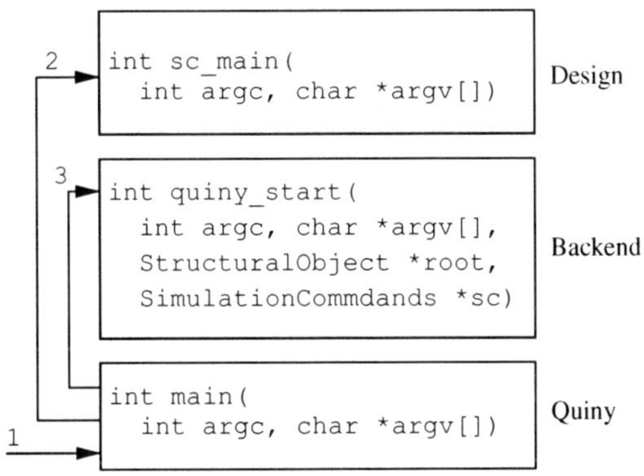

Figure 6.4. Control flow within QUINY.

```
Statement              Expression
├ IfStmt               ├ Operation
├ WhileStmt            ├── UnaryOp
└ ForStmt              └── BinaryOp

StructuralObject       Type
├ Module               ├ SimpleType
├ Port                 ├ SystemCType
└ Signal               └ ClassType
```

Figure 6.5. Excerpt of QUINY's IR class hierarchy.

Remarks

The techniques we use bear some similarity with reflection [10]. Essentially, reflection is a technique which enables a program to figure out or even modify its own structure to perform its intended operation. The main difference of our approach to classical reflection is that it can be considered to be destructive. That is, the program loses its original functionality for the purpose of finding out its structure.

An alternative technique for deconstructing expressions are expression templates proposed by [14]. As opposed to our approach which is performed at run-time, expression templates are a compile-time technique and mainly intended for optimisation.

5. Evaluation

In order to evaluate our approach with respect to the requirements formulated in Section 2.0, we implemented the ideas presented in Section 4 in a proof of concept prototype.

Currently, it is able to process the simple counter shown in Fig. 6.6 including the stimulus module shown in Fig. 6.7 and the surrounding `sc_main` function (not shown).

A prototypical VHDL generating backend was implemented. The backend in conjunction with the QUINY core library realises the exemplary SystemC to VHDL translation tool. It creates syntactically correct VHDL code, which produces signal traces identical to those of the SystemC version. The resulting VHDL code of the counter and the stimulus module is also shown in Figs. 6.6 and 6.7. The VHDL code of the top-level module which results from the `sc_main` function is left out in this article, but can be obtained from the ICODES project home page [8].

Comparing our reflexive approach to a classical parser, which directly processes the source code in text form, we see the following advantages and disadvantages.

```
#include <systemc.h>

SC_MODULE(Counter)
{
  sc_in<bool>          pi_bClk;
  sc_in<bool>          pi_bReset;
  sc_in<bool>          pi_bUpDown;
  sc_out<sc_uint<8> >  po_Count;

  sc_signal<sc_uint<8> > m_Count;

  SC_CTOR(Counter)
  : pi_bClk("pi_bClk")
  , pi_bReset("pi_bReset")
  , pi_bUpDown("pi_bUpDown")
  , po_Count("po_Count")
  , m_Count("m_Count")
  {
    SC_METHOD(main);
    sensitive << pi_bClk.pos();

    SC_METHOD(updateOutput);
    sensitive << m_Count;
  }

void main()
{
 if (pi_bReset.read() == true)
 {
  m_Count.write(sc_uint<8>(0));
 }
 else
 {
  if (pi_bUpDown.read() == false)
  {
   m_Count.write(m_Count.read() +
                 sc_uint<8>(1));
  }
  else
  {
   m_Count.write(m_Count.read() -
                 sc_uint<8>(1));
  }
 }
}

  void updateOutput()
  {
    po_Count.write(m_Count.read());
  }
};
```

```
LIBRARY ieee;
USE ieee.std_logic_1164.all;
USE ieee.std_logic_unsigned.all;
USE ieee.std_logic_arith.all;

ENTITY Counter_c_entity IS
 PORT (
  pi_bClk    : IN BOOLEAN;
  pi_bReset  : IN BOOLEAN;
  pi_bUpDown : IN BOOLEAN;
  po_Count :
   OUT STD_LOGIC_VECTOR(7 DOWNTO 0)
 );
END Counter_c_entity;


ARCHITECTURE behaviour OF
             Counter_c_entity IS

  SIGNAL m_Count :
    STD_LOGIC_VECTOR(7 DOWNTO 0);

BEGIN


main : PROCESS(pi_bClk)
BEGIN
 IF pi_bClk'EVENT  AND
  pi_bClk = TRUE THEN

  IF (pi_bReset=TRUE) THEN
   m_Count <=
     CONV_STD_LOGIC_VECTOR(0,8);
  ELSE
   IF (pi_bUpDown=FALSE) THEN
    m_Count <= m_Count+
      CONV_STD_LOGIC_VECTOR(1,8);
   ELSE
    m_Count <= m_Count-
      CONV_STD_LOGIC_VECTOR(1,8);
   END IF;
  END IF;
 END IF;
END PROCESS main;


  updateOutput : PROCESS(m_Count)
  BEGIN
    po_Count <= m_Count;
  END PROCESS updateOutput;
END behaviour;
```

Figure 6.6. SystemC implementation of an up/down counter and the output generated by the exemplary VHDL to SystemC translation backend.

```
#include <systemc.h>

SC_MODULE(Stimulus)
{
  sc_out<bool>        po_bClk;
  sc_out<bool>        po_bReset;
  sc_out<bool>        po_bUpDown;

  SC_CTOR(Stimulus)
  : po_bClk("po_bClk")
  , po_bReset("po_bReset")
  , po_bUpDown("po_bUpDown")
  {
    SC_THREAD(clockProcess);
    SC_THREAD(resetProcess);
    SC_THREAD(upDown);
  }

  void clockProcess()
  {
    while(true)
    {
      po_bClk.write(true);
      wait(sc_time(10, SC_NS));
      po_bClk.write(false);
      wait(sc_time(10, SC_NS));
    }
  }

  void resetProcess()
  {
    po_bReset.write(true);
    wait(sc_time(50, SC_NS));
    po_bReset.write(false);
    wait(sc_time(100000, SC_NS));
  }

  void upDown()
  {
    while(true)
    {
      po_bUpDown.write(false);
      wait(sc_time(100, SC_NS));
      po_bUpDown.write(true);
      wait(sc_time(80, SC_NS));
    }
  }
};
```

```
LIBRARY ieee;
USE ieee.std_logic_1164.all;
USE ieee.std_logic_unsigned.all;
USE ieee.std_logic_arith.all;

ENTITY Stimulus_s_entity IS
  PORT (
    po_bClk    : OUT BOOLEAN;
    po_bReset  : OUT BOOLEAN;
    po_bUpDown : OUT BOOLEAN
  );
END Stimulus_s_entity;


ARCHITECTURE behaviour OF
             Stimulus_s_entity IS

BEGIN

  clockProcess : PROCESS
  BEGIN
    WHILE TRUE LOOP
      po_bClk <= TRUE;
      WAIT FOR 10 ns;
      po_bClk <= FALSE;
      WAIT FOR 10 ns;
    END LOOP;
  END PROCESS clockProcess;

  resetProcess : PROCESS
  BEGIN
    po_bReset <= TRUE;
    WAIT FOR 50 ns;
    po_bReset <= FALSE;
    WAIT FOR 100000 ns;
  END PROCESS resetProcess;

  upDown : PROCESS
  BEGIN
    WHILE TRUE LOOP
      po_bUpDown <= FALSE;
      WAIT FOR 100 ns;
      po_bUpDown <= TRUE;
      WAIT FOR 80 ns;
    END LOOP;
  END PROCESS upDown;

END behaviour;
```

Figure 6.7. SystemC implementation of the stimulus module for the up/down counter (Fig. 6.6) and the output generated by the exemplary VHDL to SystemC translation backend.

Language subset. Concerning the supported language subset our current prototypical implementation is quite limited and not a real alternative to a parser. However, we see no fundamental limitations of our approach in unintrusive mode over parsers which are dedicated to finding out the design hierarchy or processing low-level SystemC models. In some aspects we expect our approach to be more robust than parsers which treat SystemC as *language*. For example, modules are recognised by the fact that they are derived from `sc_module` as opposed to detecting the convenience macro `SC_MODULE`. In this respect, our approach does not impose more constraints on the user's model than the SystemC standard does. Another example is designs where the structure is parameterised, and the number of created submodules depends on a command-line parameter or is read from a file.

Intrusiveness. For higher-level SystemC models which also utilise user-defined class types, our approach requires user intervention due to the lack of built-in reflection capabilities of C++. Another way to look at it is to say we can trade off unintrusiveness for a larger language subset. However, an SCV-like extension-mechanism [13] could be considered as "semi-intrusive". The user does not have to modify the original code, but has to provide additional code with the extensions.

Modularity. We consider the modularity aspect to be independent of the parsing approach as it is primarily a design decision concerning the implementation of the front end. For our approach we have shown in Section 4.0 how the build-up phase which heavily interferes with the user's code can effectively be decoupled from the access to the IR.

Error messages. Regarding the generation of error messages we have to distinguish two cases. The first case is the compilation of an ill-formed C++ program in conjunction with QUINY. In this case our approach is inherently inferior to a real parser. The user ends up with error messages referencing QUINY-internals. Hence, the error messages do not refer to the real source of the problem from the user's point of view. Basically, the same problem occurs when compiling SystemC models.

The second case is the compilation of well-formed C++ programs which do violate certain restrictions. For example, OSCI's standard SystemC simulation kernel reports unbound ports at run-time. Similarly, a QUINY-based synthesis backend could report a synthesis subset violation if the design contains dynamic memory allocation.

Efficiency. For a fair comparison of the efficiency, i.e. the performance of the parser, we have to take into account, that QUINY requires a complete compilation of the user code before the "parse" process – in our case the model execution – begins. In our simple example design (the aforementioned counter), the model execution including VHDL generation is negligible compared to the preceding compilation phase: On a 3 GHz Linux PC using gcc, compilation in SystemC mode took approximately 4 s, compilation in QUINY mode approximately 3 s, and model execution less than could be reliably measured with the `time` command (<0.02 s). Although the QUINY headers are not yet fully completed, we expect the final compilation times still to be slightly shorter than in SystemC mode, because QUINY's SystemC headers are less complex than the original ones, hence implying less work for the compiler.

A further interesting point is the fact that QUINY indirectly supports the compilation of separate translation units, as long as the host compiler does. Hence, it is not necessary to recompile the whole design each time a small change was made.

To conclude the efficiency aspect, we can say, that we are roughly as fast as the host compiler.

Costs. Currently we are not able to quantify the development costs and compare them to the development costs of a real parser, since (a) QUINY is still in a prototypical stage and (b) the development effort of the other approaches is rarely published.

Looking at QUINY qualitatively, we can say that we offload almost all of the complex parsing and processing to the host compiler, e.g. type checking, namespace and overload resolution, operator precedences in complex expressions, template instantiations, and so on. Harnessing the host compiler's capabilities comes at the price of finding tricks to make a C++ program find out its structure at run-time and providing replacements for all SystemC classes.

6. Conclusion

In this article we presented the first fully reflexive approach to parsing SystemC designs down to statement and expression level. We presented techniques to obtain the missing information about processes bodies at run-time. We sketched our implementation called QUINY, including an exemplary application, both of which can be downloaded from the ICODES project home page [8]. Finally, we discussed its pros and cons compared to its existing approaches.

Notes

1. The QUINY-internal code examples should be considered as pseudo-code illustrating the underlying principles. The actual implementation is a little bit more complex.

2. Actually it could also be an empty statement, which does not make much sense, but which is legal.

3. An idea to overcome this problem is to let the executable read its own debug information and try to deduce the name from the debug information. The disadvantage of this approach is that it is platform and compiler dependent.

References

[1] Berner, D., Talpin, J.-P., Patel, H., Mathaikutty, D.A., and Shukla, S. (2005). SystemCXML: an extensible SystemC front end using XML. In: *Forum on Specification & Design Languages*, pp. 405–408.

[2] Edison Design Group. Compiler front ends. `http://www.edg.com`.

[3] Fey, G., Große, D., Cassens, T., Genz, C., Warode, T., and Drechsler, R. (2004). ParSyC: an efficient SystemC parser. In: *Synthesis And System Integration of Mixed Information Technologies*.

[4] FZI Forschungszentrum Informatik, Department of microelectronic System Design. KaSCPar – Karlsruhe SystemC Parser Suite. `http://www.fzi.de/sim/kascpar.html`.

[5] Grimpe, E., Nebel, W., Oppenheimer, F., and Schubert, T. (2003). Object-oriented hardware design and synthesis based on SystemC 2.0. In: *SystemC: Methodologies and Applications*, pp. 217–246. Kluwer Academic Publishers.

[6] Hofstadter, D.R. (1991). *Gödel, Escher, Bach ein Endloses Geflochtenes Band*. dtv.

[7] Huerta, P. and Castillo, J. (Feb. 2005). *SystemC 2 Verilog Rev. 1.3*. `http://opencores.gds.tuwien.ac.at/projects.cgi/web/sc2v/overview`.

[8] ICODES. Project Homepage. `http://icodes.offis.de/`.

[9] IEEE Std 1666–2005, IEEE Standard SystemC Language Reference Manual (March 31, 2006). IEEE Computer Society.

[10] Malenfant, J. and Demers, F.-N. (1996). A tutorial on behavioral reflection and its implementation. In: *Proceedings of Reflection*, pp. 1–20.

[11] Moy, M., Maraninchi, F., and Maillet-Contoz, L. (2005). Pinapa: an extraction tool for SystemC descriptions of Systems-on-a-Chip. In: *EMSOFT '05*, pp. 317–324. ACM Press.

[12] Snyder, W. SystemPerl homepage. `http://www.veripool.com/systemperl.html`.

[13] SystemC Verification Working Group (May 2003). *SystemC Verification Standard Specification, Version 1.0e*.

[14] Veldhuizen, T. (1995). Expression templates. *C++ Report*, 7(5).

Chapter 7

MINING METADATA FROM SYSTEMC IP LIBRARY

Deepak A. Mathaikutty and Sandeep K. Shukla
FERMAT Lab, Virginia Tech, Blacksburg, USA
mathaikutty@vt.edu; shukla@vt.edu

Abstract Exploring the design space when constructing a system is vital to realize a well-performing design. Design complexity has made building high-level system models to explore the design space an essential but time-consuming and tedious part of the system design. Reduction in design time and acceleration of design exploration can be provided through reusing IP-cores to construct system models. As a result, it is common to have high-level SoC design flow based on IP library promoting reuse. However, the success of these would be dependent on how introspection and reflection capability is provided as well as the interoperability standard are defined. This leads to the important question of what kind of metadata on these IPs must be available to allow CAD tools to effectively maneuver these designs as well as allows for a seamless integration and exchange flow between tools and design methodologies. In this chapter, we describe our tool and methodology that allow introspection of SystemC design, so that the extracted metadata enables IP composition. We discuss the issues related to extraction of metadata from IPs specified in an expressive language such as C++ and show how our methodology combines C++ and XML parsers and data structures to achieve the above.

Keywords Metadata, metamodeling, component composition framework, IP composition, reflection, introspection and interoperability

1. Introduction

Metadata is defined as "data about data" and is the kind of information that describes the characteristics of a program or a model. Information ranging from the structure of a program in terms of the objects contained, their attributes,

S.A. Huss (ed.), Advances in Design and Specification Languages for Embedded Systems – Selected Contributions from FDL'06, 111–129.

methods, and properties describing data-handling details can be exemplified as metadata. This class of information is necessary for CAD tools to manipulate the intellectual properties (IPs) within system level design frameworks as well as to facilitate the exchange of IPs between tools and SoC design flows. For such exchange they express integration requirements, performance, and configuration capabilities.

We classify EDA related metadata into two broad categories: (i) interoperability metadata and (ii) introspection metadata. Interoperability metadata enables easier integration of semiconductor IP and IP tools as shown by the consortium named SPIRIT [9], which allows design flow integration by utilizing metadata exchanged in a design-language neutral format. Consider the SPIRIT enabled flow from the ARM RealView SoC Designer tool to coreAssembler, the Synopsys SoC environment in [6]. The transfer is illustrated with a ARM1176JZ-STM processor subsystem design and the metadata captured through SPIRIT include: RTL I/O signals, support for bus-interfaces, parametric configurations, abstraction levels, memory map or remap information, interconnect layout, etc. This information exported from the SoC Designer allows a seamless import of the ARM1176JZ-STM processor subsystem into the Synopsys coreAssembler. Therefore, interoperability metadata serves as a common standard for exchange of multivendor IPs between tools and design flows.

Introspective metadata on designs in system level design frameworks allow CAD tools and algorithms to manipulate the designs as well as capture interesting properties about them. These features are also useful for debuggers, profilers, type/object browsers, design analyzers, scheme generators, composition validation, type checking, compatibility checking, etc. Introspection is the ability of an executable system to query internal descriptions of itself through some reflective mechanism. The reflection mechanism exposes the structural and run-time characteristics of the system and stores it in a data structure. The data stored in this data structure is called the metadata.

Consider a component composition framework (CCF) such as MCF [3]. MCF employs introspective metadata reflected from SystemC IPs to analyze and select implementations that can be used to create an executable for the abstract specification. The metadata that would facilitate IP selection and composition includes: register-transfer level (RTL) ports & datatypes, transaction-level (TL) interface signatures, hierarchy information, polymorphic characterization, etc.

In order to reflect metadata from IPs, the important questions that need to be answered are: "what is the metadata of interest?" and "how accessible is it?". The first question would depend on how the extracted information will be used. The second question depends on how the metadata is specified. If the information is given through annotations to the implementation such as comments and pragmas, then the reflective mechanism would easily extract these.

In most cases, the information of interest is not available in a straightforward manner and therefore, extraction techniques would require mining the design to infer the metadata. This is a much harder problem and through this chapter, we show some of the challenges involved in mining metadata from designs written in a highly expressible language such as C++. Our work mainly focuses on SystemC and the metadata of interest would be used by a CCF for IP selection and composition.

2. Related Work

Several tools may be used for implementing reflection in SystemC. Some of these are SystemPerl [8], EDG [4], SystemCXML [1], or C++ as in the Pinapa [7], and KarSCPar [5].

SystemPerl [8] requires the user to provide annotations into the source file and it yields all SystemC structural information. EDG [4] is a commercial front-end parser for C/C++ that parses C/C++ into a data structure, which can then be used to interpret SystemC constructs.

SystemCXML [1] uses an XML-based approach to extract structural information from SystemC models, which can be easily exploited by back end passes for analysis, visualization, and other structural analysis purposes. It uses the documentation system Doxygen [2] and an open source XML parser [10] to achieve the structural extraction. Pinapa [7] is an open source SystemC front-end that uses GCC's front end to parse all C++ constructs and infer the structural information of the SystemC model by executing the elaboration phase. However, Pinapa is a very intrusive approach. It requires modifications of the GCC source code which makes it dependent on changes in the GCC codebase.

KaSCPar [5] is a SystemC parser that consist of two components for generating either the abstract syntax tree (AST) or a description of the SystemC design, both in XML. The output of the first tool (SC2AST) is an AST that contains a token for each C/C++ and SystemC construct in XML. These tokens capture meta-information about the construct into attributes of the token. These tokens are generated by parsing the output of the preprocessor of GNU gcc. Our work uses this tool to parse SystemC and we show the meta-information extracted is insufficient for composition. Therefore, we further mine the information to infer metadata that are lost in the high-level specification capability of the langauge.

MCF [3] is a metamodeling-based component composition framework that allows designers to (i) describe components and their interactions with a semantically rich visual front end, (ii) automatically select IPs based on sound type theoretic principles, and (iii) perform constraint based checks for composability. An XML based schema is used to store and process meta-information about the IP and facilitate automatic IP selection, composition, and generation of an executable specification for an abstract specification.

3. Metadata for IP Composition

MCF performs IP composition by selecting an implementation for an abstract specification of the design. The abstract specification contains components and channels described at different abstraction levels. In order to facilitate IP composition, we need to select an appropriate implementation that can be plugged in for the abstract specification. We are given a library of SystemC design and in order to perform selection, we need to reflect composition-related metadata. The metadata can be classified into data related to components and channels.

The metadata on an RTL component contains the set of input–output (I/O) ports and clock ports. For these ports, information related to the datatype and bitwidth are essential constituents of the metadata. Therefore, it is important to correctly extract these through different mining techniques. For a component/channel at TL, it contains a set of interface ports, clock ports, and interface access methods. Therefore, the metadata would contain the function signatures that embed the function arguments and return types. For these arguments and return types, the datatype and bitwidth need to be correctly extracted. The hierarchical structure of a component is also essential for IP selection.

Furthermore, the abstract specification can describe polymorphic components/channels, therefore, we also reflect the metadata of polymorphic implementations from the SystemC library. However, the genericity of an IP can be restricted by providing constraint on the possible type instantiations. Therefore, if the SystemC IP has been annotated with type-constraints, then it is also reflected as a part of the metadata on these polymorphic implementations.

4. SystemC Metadata

SystemC is a library of C++ classes and every design is an instantiation of some of these classes. Therefore, class-based extraction is the primary task during metadata mining. The extracted classes as shown in Fig. 7.1 helps in identifying the different modules, interfaces and further which of these modules describe components and which of them implement interfaces to describe channels. In order to find this kind of information, we extract every class and the associated inheritance tree from the design.

A module is identified in a straightforward manner, either it is an *sc_module* (base_module) or a class that inherit the *sc_module* publicly (derived_module). In Fig. 7.2, we illustrate two common structural description styles for an module X. We search the associated inheritance tree looking for an **sc_module** and if

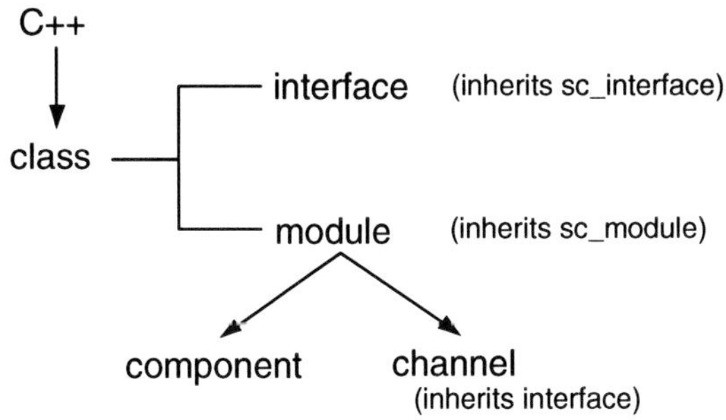

Figure 7.1. Extraction of classes.

```
// Description Styles
SC_MODULE(X) { ... };
class X : public sc_module { ... };
```

Figure 7.2. Module description styles in SystemC.

```
class CompA : public sc_behavior { ... };
```

Figure 7.3. Component description in SystemC.

one is encountered, the class gets tagged as a module. An *interface* in SystemC fulfills the following requirement:

- Must be an abstract base class (in C++)
- Inherits from **sc_interface**
- Specifies a set of access methods for a channel, but provides no implementation

Therefore, identifying *first-level interfaces* require searching the inheritance tree for an **sc_interface** and tagging them as interfaces. It is common to have interfaces that inherit first-level interfaces that we call *next-level interfaces.* This type of interface inheritance can span to any depth and therefore cannot be extracted in a single pass of the design. We implement an interface extraction procedure that recursively identifies and extract the metadata on an interface irrespective of the inheritance hierarchy.

The extraction of metadata from a module class is very different from an interface class, as their internals differ. A module contains ports and processes, whereas the interface declares function signatures. Modules are categorized into computational components and communicative channels. A *component* describes a behavior through its I/O assignments, whose implementation in SystemC is enabled with classes. It can be identified by searching for a **sc_behavior** as shown in Fig. 7.3.

A *channel* serves the purpose of a communication media and models the interaction between components. In SystemC, a channel fulfills the following requirements:

- Must derive from **sc_channel**
- Must derive from one (or more) classes derived from **sc_interface**
- Provide implementations for all pure virtual functions defined in its parent interfaces

The first requirement helps in structurally identifying a channel, which boils down to searching for **sc_channel** in the inheritance tree of the class. The second requirement helps in figuring out the list of interfaces this channel implements. The last requirement is very important as it helps in extracting the access methods of the identified channel, which is essentially the metadata captured for a channel.

As illustrated in Fig. 7.4, we have two interfaces namely *read_if* and *write_if* and a channel *fifo* that implements the interfaces. The reason for categorizing modules into components and channels is shown in Fig. 7.5, which allows us to structurally distinguish the different modules.

However, it is common practice so simply inherited from the sc_module to describe both components and channels, in this case we structurally distinguish the two under the following requirement:

- Requirement 1: A channel derives from one (or more) class(es) derived from sc_interface and a component does not.

```
// Interface description
class write_if : public sc_interface {};
class read_if  : public sc_interface {};

//Channel description
class fifo : public sc_channel,
             public write_if,
             public read_if
{ ... };
```

Figure 7.4. Interface and channel description in SystemC.

```
// Channels
typedef sc_module sc_channel;
//Components
typedef sc_module sc_behavior;
```

Figure 7.5. Module classification.

Another reason for categorizing modules is that the internals of a channel and component differ and mandates the parser to perform different extractions.

Metadata on interfaces. In order to avoid problems with multiple inheritance and provide safe interfaces in SystemC, interfaces never implement methods or contain data members. Given the way of describing of an interface in SystemC, the only metadata of interest is function signatures of access methods as shown in Fig. 7.6. To obtain a function signature, the parser extracts the return type and function arguments of the method in terms of their name, datatype, and bitwidth.

Metadata on components. A component in SystemC is a collection of ports and processes. Ports play a crucial role in describing the structure of a component, whereas processes describe the computation. Therefore, the main task while mining metadata of a component is to correctly extract its ports and port-level details. The ports can be classified into two types based on the abstraction level of the component. A component could be described at the RTL level of abstraction, which would require it to have I/O ports. It can also be described at the transaction level, where it communicates through interface-level ports. A component's port-level RTL description is facilitated through the constructs such as **sc_in**, **sc_out**, and **sc_inout** in SystemC as shown in Fig. 7.7. A component's interface-level TL description is facilitated through the construct **sc_port** construct in SystemC, which gives the component access to the methods of the interface. Similar to the I/O ports, we extract the clock port, which is provided using the **sc_in_clk** and **sc_out_clk** constructs in SystemC.

```
//1st level interface
class simple_bus_blocking_if : public sc_interface {
// Interface access method
virtual void read(unsigned int unique_priority, int *data,
unsigned int address, bool lock) = 0; ... }
```

Figure 7.6. Interface access method *read* in SystemC.

```
// Input port
sc_in<sc_int<16> > in_real;
// Interface port
sc_port<read_if> data_in;
```

Figure 7.7. Ports in SystemC.

SystemC allows for structural hierarchy by instantiating modules within a module. The metadata on hierarchical structure of a component boils down to its module hierarchy.

Metadata on channels. The main purpose of channels in SystemC is to facilitate transaction-level modeling, where the communication semantics is rendered through the interface-level function calls. As a result while mining for metadata on a channel, we extract its interface-level function declarations, interface ports and clock ports to describe its structure. Looking at the interfaces that a channel inherits, we can separate the interface access methods from all the function declarations.

It is commonly seen as an industry practice to avoid the usage of sc_port construct during TL modeling (TLM) due to issues related to synthesis. However, to facilitate TLM, it is common to have a component inherit a channel implementing the interface. This would violate our first requirement that makes it possible to distinguish a component from a channel. Therefore, we make an addendum to the requirement.

- Requirement 2: If a component derives from one or more classes (besides the **sc_module**) then these classes should be channels.

Metadata on polymorphic components & channels. In C++-based HDLs such as SystemC, a component/channel described using a partially specified C++ template is polymorphic and uses type variables as place holders for concrete types that are resolved during instantiation. Some examples of such implementation are buffers, fifos, memories, and switches (Fig. 7.8).

In order to capture metadata on these polymorphic descriptions we start by extracting the template expression. Then for the implementation, the polymorphic parts are the ports or interface functions. Examples of such an I/O and interface port is shown in Fig. 7.9. Therefore, we need to extract the polymorphic nature of these as a part of the metadata of ports.

For an interface function, the polymorphic nature can be attributed to function arguments or return type (Fig. 7.10) and is dealt with in a similar manner as polymorphic ports. Another place, where the effect of templates need to be taken into account is during hierarchical embedding. If the submodules of a module are templated, then during the extraction of the metadata on structural hierarchy, the polymorphic nature of submodules should be captured.

```
//Polymorphic component description
template < class DataT, unsigned int size >
class fifo : public sc_module { ... };
```

Figure 7.8. Generic FIFO.

```
//Polymorphic I/O port
sc_in<DataT> data_in;
//Polymorphic interface port
sc_port<read_if<DataT> > data_in;
```

Figure 7.9. Polymorphic ports.

```
//Polymorphic function argument
void read(DataT &);
//Polymorphic return type
DataT & write();
```

Figure 7.10. Polymorphic interface functions.

Metadata from annotations. Designer can explicitly annotate the code with metadata and it is commonly carried out through comments or pragmas. We allow the designer to insert comment-level annotations into the implementation, that would guide the constraining aspect of the polymorphic component/channel. The user inserts hints through comments of a certain format into the code from which we obtain the metadata used for constraining a generic implementation. A comment that specifies a constraint on some component/channel has the format shown below:

constraint : *name* $< arg_1, \ldots, arg_n >$, where $arg_i = type_1; type_2; \ldots$

The comment-level constraint has a format similar to the template expression. The *name* describes the component on which the constraint is enforced. arg_i has a one-to-one correspondence to a template argument based on its position in the template expression. However, arg_i captures legal C++ or SystemC types that could be instantiated instead of the place holder in the template expression. The designer inserts the specialized comments into the pertaining IPs. These are extracted as a part of the metadata on the component/channel on which the constraints are specified.

Mining metadata. We extract fragments of the implementation, which are language-level specifics that are processed to guide the extraction of certain metadata that not available in a straightforward manner. The language-level specific are indirections that ease the designer's implementation process. C++ provides many high-level constructs (indirections) to eliminate the repetition or tediousness associated with the usage of other level-constructs. These indirections hide metadata from the straightforward extraction of the implementation. Therefore, we extract these language-level specifics and process these to reveal the hidden metadata.

An example in C++ is name aliasing introduced through constructs such as *typedef*. The usage is to insert a simple alias for a long statement that is used in multiple places in the implementation. We call the usage of these constructs as *type indirection*, since they hide the actual type of a port or interface during reflection. As a result, we perform a search & replace of the alias with the actual legal type. This requires identifying a name alias, the actual type associated with the alias and then updating the tokens with the actual type. Therefore, we extract are the usage of the following access modifiers *#define*, *typedef*, and *const*. The need to extract these constructs such that they can be processed in search of any indirection is not trivial. We achieve this by implementing a replacement procedure that incrementally replaces the *type indirection* to arrive at the actual datatype.

5. Tools and Methodology

Given a library of SystemC IPs, we achieve automated extraction of metadata from these designs using tools such as KarSCPar and Xerces-C++ parsers and a methodology built on top of two languages: C++ and XML. The objective is to facilitate IP selection and composition by the CCF making use of the extracted metadata from the IP library. The methodology has different stages that extract, infer and constraint the metadata from the given library of IPs. The design flow for our methodology is shown in Fig. 7.11. It has three stage, which begins with a library of SystemC designs and ends at an XML document object model which embeds the metadata of these designs. It is called the component DOM (cDOM) and serves as a primary input to the CCF. In the following subsections, we elaborate on the various stages.

Stage 1: SystemC Parsing

The input to this stage is designs written in SystemC and compiled using a C++ compiler. We use the KaSCPar SystemC parser to traverse the designs and print out the corresponding XML, which is an AST. It contains some meta

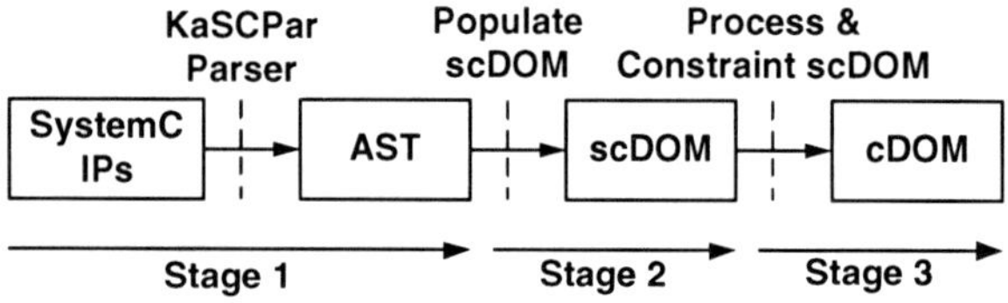

Figure 7.11. Design flow of the methodology.

information like column number, line number, and file name as a reference to the original SystemC file and are saved as attributes of each AST token. The comments of source files are also saved as special token comment in the AST.

In this stage, we obtain a more generic C++-based extraction than a specific SystemC metadata reflection that can be geared for CCF usage. As a result, the AST is too detailed and contains irrelevant information. Furthermore, to reflect certain metadata of a design, we need to process the first-level extract and perform analysis. This analysis tries to decipher the indirection caused by high-level constructs of the language. Therefore, all further processing is based on the AST.

Stage 2: AST Parsing & sc_DOM Population

In this stage, we parse the AST using the Xerces-C++ parser [10] to extract a subset of meta-information necessary for IP composition and populate the internal data structure of the parser. The data structure is a Document Object Model (DOM) that serves as a platform- and language-neutral interface that will allow programs and scripts to dynamically access and update the content of the DOM. Therefore, attaching the DOM with an introspective architecture that implements the querying interface can be achieved with ease. The output of this stage is called the **sc_DOM**, which is given as input to the last stage. We briefly describe the subset of the AST extracted to populate the **sc_DOM** below.

The primary task as explained in the previous section is the extraction of classes, which is categorized into components, channels, and interfaces. The class information is extracted from the AST and used to populate the **sc_DOM** through a **class_specifier** token. This token has an internal structure that captures the data members and member functions associated with the class. It also captures the inheritance tree of this class, which is essential for the categorization of the SystemC modules. We illustrate some of the tokens and the metadata captured during the population of the **sc_DOM**.

Interfaces. The metadata captured for an interface is the function signatures. For the simple_bus_blocking_if read access method shown in Fig. 7.6, the metadata is captured by a **function_declaration** token shown in Fig. 7.12 and wrapped in an **interface** token.

Components. The metadata captured for a component is the port-level specifics. The parser generates a **port_declaration** token for each port encountered in the component, which embeds the name, datatype and bitwidth of the port in its attributes. The extraction of an interface port for a component at TL, is performed very similar except that the datatype attribute of the **port_declaration**

```
<interface name="simple_bus_blocking_if">
    <function_declaration name="read" partial_spec="false"
    return_type="void">
        <argument datatype="unsigned int" name="unique_priority"
        partial_spec="false"/>
        <argument datatype="int" name="data"
        partial_spec="false" ptr="true"/>
        <argument datatype="unsigned int" name="address"
        partial_spec="false"/>
        <argument datatype="bool" name="lock"
        partial_spec="false"/>
    </function_declaration>
</interface>
```

Figure 7.12. **function_declaration** token *read* in XML.

```
// Input port
<ports_declaration IOtype="sc_in" bitwidth="16" datatype="sc_int"
name="in_real" partial_spec="false"/>
// Interface port
<ports_declaration IOtype="sc_port" datatype="read_if"
name="data_in" partial_spec="false"/>
```

Figure 7.13. **ports_declaration** token in XML.

captures the interface name. In Fig. 7.13, we illustrate the **port_declaration** token for the input port *in_real* and interface port *data_in* shown in Fig. 7.7.

Channels. The metadata captured for a channel pertain to port-level specifics and function signatures for the implemented access methods. For each access method encountered, we generate a **function_declaration** token similar to Fig. 7.12 and for each interface port, we generate **port_declaration** as shown in Fig. 7.13.

Polymorphic components & channels. We extract the template expression that describes a polymorphic entity and express its arguments as **template_parameter** tokens. These tokens capture three essential details about any argument besides the datatype and name. The attributes of a **template_parameter** token that embeds these details are:

1. *partial_spec* – Determines whether the argument is partially specified.

2. *has_constraint* – Determines whether the partially specified attribute is type constrained.

3. *position* – Captures the argument position in the template expression, which acts as the key for the processing and constraining performed on the component in stage 3.

In Fig. 7.14, we show the **template_parameter** tokens used to capture the essentials of each argument in the template expression on the fifo component in Fig. 7.8. The following observations about the fifo can be made from this example: (i) fifo is polymorphic w.r.t the data it stores (DataT & partial_spec="true") and (ii) it is a generic implementation with no specified constraints (has_constraint="false").

During extraction of polymorphic ports, the **port_declaration** token captures the template argument as its datatype and uses the partial_spec attribute to state that the port transmits/recieves a template type. Extracting interface functions requires worrying about polymorphic function arguments or return types (Fig. 7.10) and is dealt with in a similar manner as polymorphic ports. If the submodules of a module are templated, then during the extraction of the module hierarchy, the corresponding **template_parameter** tokens are generated.

General extraction. Here we extract, high-level constructs, which undergoes mining techniques to identity hidden metadata. Figure 7.15 illustrates the usage of *typedef* and its token representation in XML, where the attribute access_modifier captures the type of construct.

Comments. The corresponding XML generated is called the **constraint_entry** token. It consists of a set of **argument** tokens and identifies with the template arguments through an attribute that captures its position in the template expression. For each **argument**, we insert a **dt_entry** token that captures the

```
<component name="fifo">
<template_parameter has_constraint="false" datatype="class"
name="DataT" partial_spec="true" position="1">
<template_parameter datatype="unsigned int"
name="size" partial_spec="false" position="2"/>
</component>
```

Figure 7.14. **Template_parameter** token.

```
// High-level construct
typedef sc_int<16> DataT;
// Type_declaration token
<type_declaration access_modifier="typedef" bitwidth="16"
data_type="sc_int" name="DataT" template_type="false"/>
```

Figure 7.15. **Type_declaration** token for *typedef*.

```
//constraint : fifo < int; sc_int<32>; sc_int<64>, NA >
<constraint_entry class="fifo">
    <argument position="1">
        <dt_entry type="int"/>
        <dt_entry length="32" type="sc_int"/>
        <dt_entry length="64" type="sc_int"/>
    </argument>
</constraint_entry>
```

Figure 7.16. Illustration of comment-level constraint & corresponding **constraint_entry** token.

datatype and bitwidth for all the legal types allowed through the constraint. Note that in Fig. 7.16, the second argument of the template expression is fully specified and is mapped to 'NA' in the constraint implying that it has no bearing on the second argument.

Stage 3: Processing & Constraining sc_DOM

This is a two phase stage with the populated **sc_DOM** given as input. The **sc_DOM** can be abstractly seen as a collection of components, channels, and interfaces. In this stage as shown in Fig. 7.17, some of the **sc_DOM** constituents undergo a processing that results in updating/modifying some of the tokens. The processing is followed by a constraining phase where the stored implementations are type-restricted by propagating the effect of the specified constraint on the generic aspects of the implementation. The output of this stage IP Library DOM called **cDOM**.

Phase 1: Processing sc_DOM In this phase, we identify some ignored information from the AST parsing stage that requires analyzing the contents of the sc_DOM. Secondly, we process the sc_DOM to remove any indirection that hides the structural aspects of the design through place holders. The discarded information that try to revive in this phase pertain to next-level interfaces. As mentioned in stage 2, we only identify first-level interfaces, however, it is common to have next-level interfaces. Figure 7.18 shows a second-level interface *rdwr_if* that inherit first-level interfaces *read_if* and *write_if* of Fig. 7.4.

In stage 3, when we encounter a class element during the parsing of the AST, we try to identify it as a second interface. If successful, then the corresponding **interface** token is generated and appended to the sc_DOM. It is necessary to identify both first- and second-level interfaces, for the correct extraction of the TL behavior of a channel. The reason being, one of the key step in the extraction of a channel is to separate interface access methods from internal function definitions, which would require knowledge to all the interfaces that this channel implements. Therefore, after the extraction of the second-level

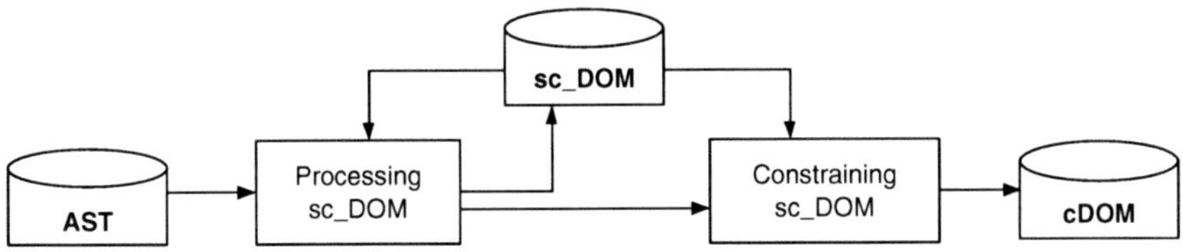

Figure 7.17. Stage 3: Processing & constraining sc_DOM.

```
//Inherits 1st level interfaces read_if and write_if
class rdwr_if  : public read_if, public write_if { ... }
```

Figure 7.18. Second-level interfaces.

interfaces, we perform another round of extraction for the channels that implement this interface. This results in addition of more **interface_function** tokens to the **channel** token.

An interesting point to note is that second-level interfaces can give rise to third-level interfaces, and so on. This unidirectional chain can continue to any depth and therefore we implement an extraction procedure shown below, which helps in navigating the chain and extracting the interfaces correctly. SystemC only allows for interfaces through safe usage of multiple inheritance, therefore no cyclic dependency exist in the chain. Therefore, we implement the algorithm shown below to correctly extract the interface hierarchy.

Interface extraction

{Given **C** set of all classes and **IF** = **OC** = ϕ}
Step 1 For each $c \in$ **C**,
  Step 1.1 If c inherits *sc_interface* then insert(**IF**,c) else insert(**OC**,c)
Step 2 For each $oc \in$ **OC**
  Step 2.1 If oc inherits an interface *intf* s.t. $intf \in$ **IF** $\wedge\ oc \notin$ **IF**,
    then insert(**IF**,oc)

Consider the following code snippet in Fig. 7.19, the tokens shown are generated at the end of stage 2. The datatype extracted for the port is *DataT*, which basically is an alias for the actual type. Therefore, to update such incorrect extractions, we apply the type-replacement procedure on the sc_DOM that removes all type indirections. Some of the structural entities that are affected by type indirection are:

- I/O and interface ports
- Interface function arguments and return types

```
//Type indirection
typedef sc_int<16> DataT;

// IO Port
sc_in<DataT> in_real;
// Type_declaration token
<type_declaration access_modifier="typedef" bitwidth="16"
data_type="sc_int" name="DataT" template_type="false"/>

// Port_declaration token
<ports_declaration IOtype="sc_in" bitwidth="16" datatype="DataT
name="in_real" partial_spec="false"/>
```

Figure 7.19. Example of type indirection & tokens generated.

- Template expression and template arguments
- Submodule instantiations

The replacement procedure runs through these entities searching for any indirection and replacing it with the actual type. This problem is further convoluted due to the same deep-chain problem associated with interfaces. The type indirection can be repeated to any level, which requires iterating through the complete chain to find the actual type. The replacement procedure takes this problem into account and through an iterative scheme manages to replace all the type indirections at any depth with their respective datatypes. The procedure is shown below:

Replacement procedure

{Given
T an ordered set of all type indirections,
P set of all IO/interface ports,
I set of all interface functions,
E set of all template expressions,
S set of all submodule instantiations}
Step 1 For each type indirection $t_i \in \mathbf{T}$,
Step 2 For each type indirection $t_j \in \mathbf{T}$,
  Step 2.1 If **search**(t_j, t_i) = *true* then **replace**(t_j, t_i)
Step 3 For each port $p \in \mathbf{P}$,
  Step 3.1 If **search**(p, t_i) = *true* then **replace**(p, t_i)
Step 4 For each interface $n \in \mathbf{I}$,
  Step 4.1 If **search**(n, t_i) = *true* then **replace**(n, t_i)

Step 5 For each template expression $e \in \mathbf{E}$,
Step 5.1 If **search**(e, t_i) = *true* then **replace**(e, t_i)
Step 6 For each submodule instant $s \in \mathbf{S}$,
Step 6.1 If **search**(s, t_i) = *true* then **replace**(s, t_i)

This simple procedure works well, because of the order in which the **type_declaration** tokens are generated. In C++ to insert a typedef instance, the statement has to be valid, therefore the order in multilevel typedef-ing is very important and necessary for this procedure to terminate. We generate tokens and populate T in the same order in which the typedefs are specified in the implementation. In Fig. 7.20, we make use of a 2D templated fifo example to illustrate the application of the replacement procedure. The outcome of this phase is the updated **sc_DOM**, which is more complete and correct in terms of metadata.

Phase 2: Constraining sc_DOM We discussed the extraction of comment-level constraints that a library engineer specifies to restrict the genericness of the IP. In this phase, these constraints captured as *constraint_entry* tokens are propagated through the IP. The propagation limits some polymorphic structural aspect of the IP to a set of legal types specified in the constraint.

In Fig. 7.16, we had illustrated a comment-level constraint on the storage type of the fifo in Fig. 7.8. The resultant after propagating the constraint is shown in Fig. 7.21, which appends to the **template_paramter** token for DataT, a set of **dt_entry** tokens that capture the possible types for substitution.

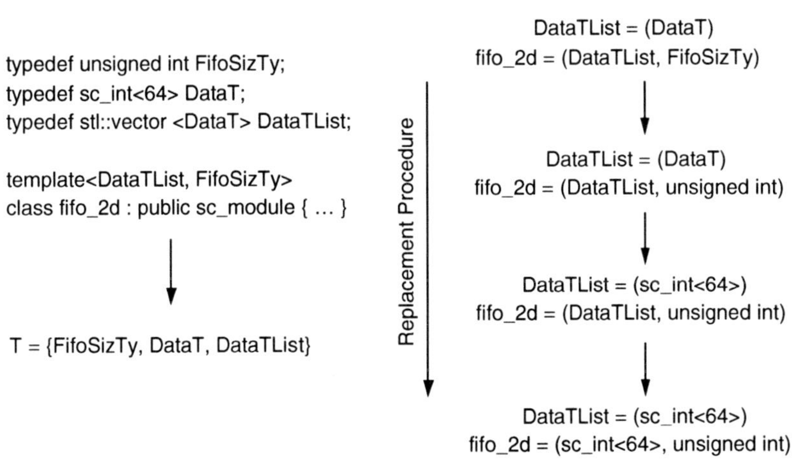

Figure 7.20. Application of the replacement procedure.

```
<template_parameter constrainted="true" datatype="class"
name="DataT" partial_spec="true" position="1">
    <dt_entry type="int"/>
    <dt_entry length="32" type="sc_int"/>
    <dt_entry length="64" type="sc_int"/>
</template_parameter>
```

Figure 7.21. Constrained **template_parameter** token.

The polymorphic aspects of an IP range from I/O ports to structural hierarchy. Therefore, the constraint propagation would require appending all of these aspects with the appropriate **dt_entry** tokens. The outcome of this phase is the **cDOM**.

6. Conclusion

We have illustrated the extraction of introspective metadata from SystemC IPs. Our methodology combines the KarSCPar SystemC Parser, XML tools, and the DOM data structure to enable reflection and introspection of SystemC designs. The metadata serves as input to a CCF which allows for IP selection and composition to create an executable given an abstract specification and a library of SystemC IPs. Furthermore, we have outlined some of the challenges in terms of multistage processing required to mine metadata for composability of IPs from designs specified in C++.

References

[1] Berner, D., Patel, H.D., Mathaikutty, D.A., and Shukla, S.K. (2005). SystemCXML: an extensible systemc frontend using XML. In: *Proc. Forum on Design and Specification Languages (FDL '05).*

[2] Doxygen Team Doxygen. http://www.stack.nl/ dimitri/doxygen/.

[3] FERMAT MCF: A Metamodeling-based Visual Component Composition Framework. http://fermat.ece.vt.edu/MCF/MCF.htm.

[4] EDG Edison Design Group C++ Front-End. Website: http://edg.com/cpp.html.

[5] KaSCPar KaSCPar – Karlsruhe SystemC Parser Suite. Website: www.fzi.de/KaSCPar.html.

[6] Grun, P., Baxter, C., Noll, M., and Madl, G. (2006). Integrating a multi-vendor esl-to-silicon design flow using spirit. *Design & Reuse, The Catalyst of Collaborative IP-Based SoC Design.*

[7] Moy, M., Maraninchi, F., and Maillet-Contoz, L. (2005). An extraction tool for SystemC descriptions of systems-on-a-chip. In: *Proc. of*

ACM International Conference on Embedded Software (EMSOFT '05), pp. 317–324.

[8] SYSTEMPERL Snyder, W. http://www.veripool.com/systemperl.html.

[9] SPIRIT SPIRIT Schema Working Group. http://www.spiritconsortium.org/.

[10] APACHE The Apache Software Foundation, Xerces C++ Validating XML Parser. Website: http://xml.apache.org/xerces-c/.

Chapter 8

NONINTRUSIVE HIGH-LEVEL SYSTEMC DEBUGGING*

Frank Rogin[1], Erhard Fehlauer[1], Steffen Rülke[1],
Sebastian Ohnewald[2], and Thomas Berndt[2]

[1]*Fraunhofer-Institut für Integrierte Schaltungen IIS*
Branch Lab Design Automation
Zeunerstr. 38, 01069 Dresden, Germany
{frank.rogin; erhard.fehlauer; steffen.ruelke}@eas.iis.fraunhofer.de

[2]*AMD Saxony LLC & Co. KG / Dresden Design Center (DDC)*
Wilschdorfer Landstr. 101, 01109 Dresden, Germany
{sebastian.ohnewald; thomas.berndt}@amd.com

Abstract We present a nonintrusive high-level SystemC debugging approach to be used with SystemC v2.0.1 and GNU debugger GDB. Our approach is integrated into an industrial design flow and enables developers to debug designs at high-level working with signals, ports, events, and processes. Thus, one gets quick and concise insight into static structure and dynamic behavior of the design without the burden of gaining detailed knowledge of the underlying SystemC simulation kernel. Only minor transparent changes to SystemC kernel source code are required, whereas there is no need to touch the flow or the designs. Practical experiences show promising results.

Keywords High-level Debugging, validation, system level design, SystemC, GDB

1. Introduction

System level design methodologies promise to address major challenges in modern System-on-Chip (SoC) designs. System level design embraces various abstraction levels, different components (IP, SW/HW), diverse tools,

*Partial funding provided by SAB-10563/1559 and European Regional Development Fund (ERDF)

S.A. Huss (ed.), Advances in Design and Specification Languages for Embedded Systems – Selected Contributions from FDL'06, 131–144.

and methodologies which further complicate design comprehension. Studies revealed that today often more than 50% of design time is spent to verify a complex design that means to identify, understand, localize, and correct errors [3].

Many system level languages and design environments were proposed over the last years, e.g. [2, 10]. One of the most popular languages of this type is SystemC [11]. It has become a de facto standard in industry and in the academic field. SystemC provides concepts such as object-orientation, concurrency, and high-level modeling.

Currently SystemC does not comprise debugging aspects. It solely defines functions to trace module level variables and signals. Traced values are written to file during simulation, and analyzed with standard tools afterwards. Standard C++ debuggers are applied to analyze a functions local variables during simulation run. Unfortunately, both debugging approaches operate on very low abstraction level. Especially, standard C++ debuggers do not understand specific SystemC constructs. Besides, SystemC maps modules onto individual threads of execution which leads to nonlinear execution sequences. This makes predicting which module will be active next extremely difficult.

As working at appropriate abstraction levels is an essential means to understand designs and to fix bugs quickly, several commercial and research tools have been developed dealing with high-level SystemC debugging. Some of the available commercial solutions and academic prototypes are listed and assessed below.

MaxSim Developer Suite [1] comprises a block level editor, and simulation, debugging, and analysis tools. It addresses architectural analysis as well as SystemC component debugging at low level and at transactional level. *ConvergenSC System Verifier* [12] targets SystemC system level design and verification. It utilizes a simulation kernel which is specially adopted to fit SystemC needs. Its integrated debugger offers SystemC specific commands supporting breakpoints and SystemC QThreads at source level. *CoCentric System Studio* [13] supports SystemC design, simulation and analysis at system level, and partly synthesis from behavioral and RT level. It utilizes standard C++ debuggers (e.g. GDB), i.e. it does not handle SystemC constructs in a specific way.

[7] presents a method to extract structural data from SystemC designs automatically, and to pass it to a commercial visualization tool using an application programming interface (API). The SystemC kernel has been modified to interface to the API. [9] uses SystemC simulation results to create Message Sequence Charts to visualize SystemC process interaction at a high level. Filters cut out parts of interprocess communication in order to reduce information complexity. [4] applies the observer pattern [8] to connect external software to the SystemC simulation kernel. This general method facilitates loose coupling and requires just minimal modifications of the SystemC kernel.

None of the tools mentioned above fully meets the requirements to integrate high-level SystemC debugging into the existing design flow at AMD, namely to

- Integrate into the flow where designers used to apply GDB either at command line or through a GUI
- Easily access static and run-time design information at system level
- Work with an existing SystemC kernel
- Avoid changes to the design to support debugging
- Exercise high-level breakpoints

So we decided to implement high-level SystemC debugging as a set of GDB user commands to avoid patching of GDB source code. C++ routines and shell scripts collect required data from SystemC or GDB run-time, respectively, and present the information to the designer. Only minor transparent changes to SystemC kernel source code were made to enhance debugging performance.

The remainder of this paper is organized as follows. Section 2 proposes our high-level SystemC debug methodology derived from the given industrial requirements. Section 3 introduces implementation details, and explains modifications made to improve performance. Section 4 presents some practical experiences gained. Section 5 concludes the paper.

2. Methodology

Requirements and Design Issues

The most important industrial requirement was the demand for a nonintrusive debugging facility that fits seamlessly into the existing design flow. That means, the solution should work with the available SystemC kernel and avoid any changes to present designs or (third-party) IP blocks. On the tool side, the already applied GNU debugger GDB [6] should be extended without any need for patching its sources. Advantages are a familiar, intuitive, and unchanged debugging flow combined with a minimal learning curve for the user. Moreover, maintenance and customization of the flow are reduced to a minimum.

Debugging at system level requires various kinds of high-level information that should be retrievable fast and easily. According to [5], one main information category is of interest in the debugging context:

Run-time infrastructure information can be divided into three subcategories. (i) *Static simulation information* describes the structure of the architecture that means the number of modules, the number of processes and signals, the I/O interfaces and their connections, etc. (ii) *Dynamic simulation information* includes among other things the triggering conditions of processes, the process sensitivity lists, and the number and types of events in the simulation queue. (iii) *Debugging callbacks* (here, *high-level breakpoints*) allow the simulation environment to break on certain events such as process activation, value changes on signals, or the ongoing simulation time.

Studies at AMD indicated several debug patterns typically used in daily work. A pattern describes the steps (in GDB) to enquire needed debugging information at system level. Based upon characteristic debug patterns, high-level commands were implemented (Table 8.1).

At top level, commands are classified in examining and controlling types. In a distributed development flow, many designers are working on different components at the same design. In case of an error, it is essential to get a fast insight into external components and their interaction with your own ones. For

Table 8.1. High-level debugging commands.

Examining commands	
Static simulation information	
lss	list all signals in given hierarchy
lsm	list all modules in given hierarchy
lse	output all events instantiated in modules
lsio	list I/O interface in given hierarchy
lsb	list all bindings of specified channel
Dynamic simulation information	
lpt	list all trigger events of all processes (w.r.t. a specific time stamp)
lst	output code line a process is currently pending
lpl	show all processes listening on given event
lsp	output all [c]thread and method processes
Controlling commands	
ebreak	break on next invocation of processes that are sensitive to specified SystemC event
rcbreak	break on next invocation of processes that are sensitive to rising edge of given clock
fcbreak	break on next invocation of processes that are sensitive to falling edge of given clock
pstep	break on next invocation of given process
dstep	break on processes which will be active in the next simulation delta cycle
tstep	break on processes which will be active in the next simulation time stamp

this reason, examination commands retrieve either static or dynamic simulation information. Controlling commands provide high-level breakpoints to reach a point of failure at system level very quickly. They stop program execution at certain conditions, such as the next activation of a specific process or all processes which are sensitive to a given SystemC event.

General Architecture

Figure 8.1 illustrates the layered architecture of our high-level debugging environment. Due to the demand for a nonintrusive extension of GDB, all high-level debugging commands are implemented on top of it. Command sequences are encapsulated as a unit in a user-defined command composing the so called **macro instruction set** at the user layer. A macro instruction implements a desired debug functionality by using built-in GDB commands (e.g. examining the symbol table, or the stack), and a set of additionally provided **auxiliary functions** at the API layer. Auxiliary functions are C++ or script helpers that evaluate and process information supplied by the **debug data pool** representing the data layer. The pool obtains its content either from redirected output of GDB commands (temporary log files), data structures of SystemC kernel classes, or a debug database holding preprocessed information collected during initialization of the actual debug session.

Debug Flow

Examining commands (Table 8.1) mostly just process directly accessible information provided by the debug data pool. In contrast, a controlling command comprises a complex interaction between data provided by the pool and GDB. Here, execution starts usually with a redirection of a GDB command

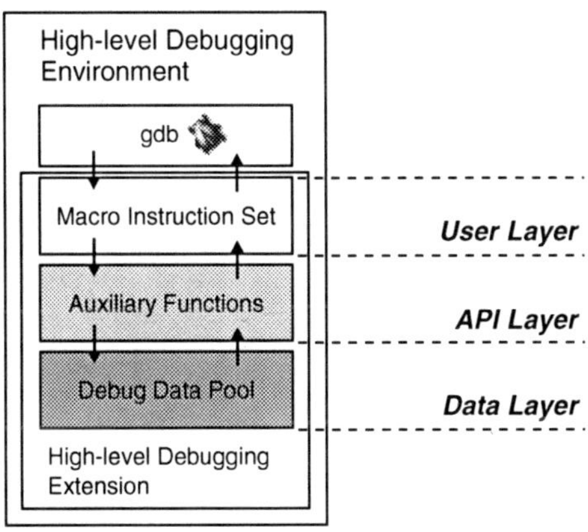

Figure 8.1. High-level debugging environment.

output (e.g. `backtrace`) into a temporary log file. The log content is evaluated by an affiliated auxiliary function. According to the specific debugging task, extracted log data trigger various actions:

- Storing or updating data into the debug database
- Caching data in temporary data structures
- Retrieving enquired debug data from the database
- Generating a temporary GDB command file

A generated temporary GDB command file is sourced subsequently. Its execution releases either an instant action or it creates a (temporary) breakpoint which is triggered in the future. According to the specific task the loop of writing and evaluating log files, and performing various actions can run some further times. As a result, data are stored in the debug database, or enquired information is output at GDB console. In addition, oncoming debugging commands and data collection actions can be prepared by caching data, or setting (temporary) breakpoints. Figure 8.2 sketches the exemplary execution of an imaginable debugging command. There, each participating component belongs to one of the three layers.

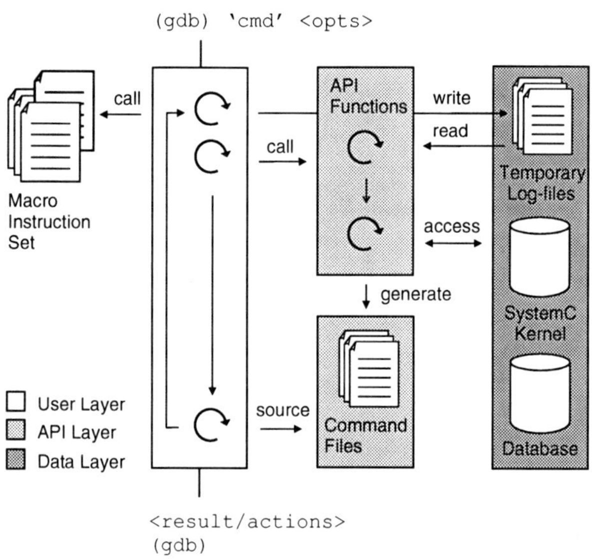

Figure 8.2. Exemplary debug flow.

3. Implementation

User Layer

The user layer acts as the interface to the high-level debugging capability. It comprises the macro instruction set which is summarized in several GDB script files. Furthermore, this layer contains GDB scripts to setup and to initialize the debugging environment.

Example. The lsb command (Table 8.1) presents the common command implementation template at the user layer.

```
define lsb
if ($hd_elaborated)
  echo ---lsb: list all bound ports---\n
  call hd::list_bound_ports($arg0)
  echo -------------------------------\n
else
  echo not elaborated yet\n
end
end
document lsb
list all bindings for the specified channel
end
```

API Layer

The API layer supports the implementation of high-level debugging commands at the user layer. It divides into an auxiliary function API and a database API. The **auxiliary function API** comprises in addition to awk/shell scripts, particularly C++ functions which realize more sophisticated helper functionality. Scripts are normally used to straightforward process text files. Implementations of the same functions showed a significant performance yield of the C++ over the script-based realization.

The **database API** supplies functionality to store data into, and to retrieve data from the debug database. For each database type a set of access functions is provided.

Example. A call of the lsb command invokes the C++ auxiliary function hd::list_bound_ports(const char) which retrieves the corresponding sc_interface instance using the SystemC method sc_simcontext::find_object(). Afterwards it calls hd::list_binding(sc_interface*). This database API function fetches the static binding information from the debug database and formats them accordingly for output (Fig. 8.3).*

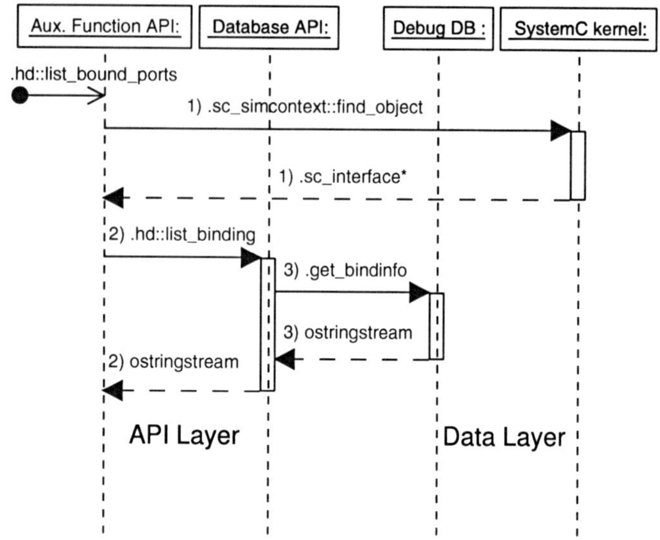

Figure 8.3. lsb command at the API layer.

Data Layer

Three data sources compose the data layer supplying either static or dynamic simulation information.

SystemC kernel. The SystemC kernel provides some basic introspection capabilities useful for retrieving design and run-time information. Various global registration classes allow to query *static simulation information*, such as port, module, channel, or SystemC object registry. For instance, the object hierarchy can be easily browsed using the following loop:

```
sc_simcontext* c = sc_get_curr_simcontext();
sc_object* o = c->first_object();
while (o) {
   if(!strcmp(o->kind(),"sc_module")) {
      // module specific actions
   }
   else if(!strcmp(o->kind(),"sc_signal")) {
      // signal specific actions
   }
   ...
   o = c->next_object();
}
```

The simulation control, implemented by the kernel class `sc_simcontext`, supplies many valuable *dynamic simulation information* such as runnable processes at the next delta cycle, or the delta event queue.

Temporary log files. Such log files will be created by redirecting the output of GDB commands (e.g. `thread`, `backtrace`) providing *dynamic simulation information* only accessible at debugger side such as the assigned GDB thread ID of a SystemC thread process.

Debug database. During setup of a new debug session, *static simulation information* is logged and stored into the debug database using GDB (at least in the first implementation approach). Here, we utilize particularly the ability of a debugger to fetch private class data in the SystemC kernel which do not have public access methods. Required debug functionality (Table 8.1) bases on four information classes: event, binding, method, and thread process information. Each class is represented by its own datatype holding preprocessed (e.g. process entry function name), kernel-private (e.g. process handle private to `sc_process_table`), or special debug session data (e.g. GDB thread ID). Figure 8.4 sketches the UML class diagram of the debug database showing only the datatypes representing the information classes together with their attributes.

Example. The following lsb call retrieves the binding information for a channel of an example application referenced by its hierarchical object name i0_count_hier.count_sig. The database API queries for the proper hd_db_bindinfo instance using the corresponding `sc_interface` *object, fetches, and formats its data.*

```
(gdb) lsb "i0_count_hier.count_sig"
---lsb: list all bound ports---
bindings of channel i0_count_hier.count_sig
Driver:
  i0_count_hier.i_counter.outp <sc_out>
Drivee:
  i0_count_hier.i_signal2fifo.inp <sc_in>
------------------------------
```

Performance Issues

Practical tests on real-world AMD applications revealed a considerable performance problem using the pure nonintrusive implementation approach. Especially, the setup phase of the extended GDB takes an unacceptably long time (Table 8.2). Investigations indicated particularly the assemblage of the high-level debugging information as the bottleneck. Here, hidden breakpoints

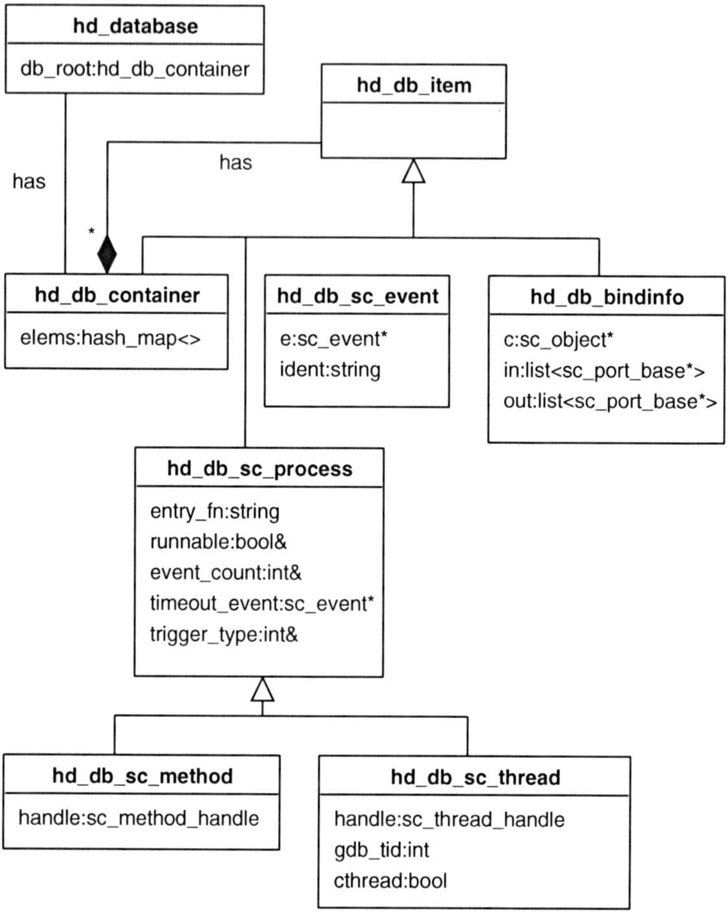

Figure 8.4. Debug database class hierarchy.

in the SystemC kernel registered and triggered actions to handle the instantiation of processes and events. So, we developed a second approach to accelerate the data assemblage phase. The idea is to reduce the number of breakpoints while moving their functionality into the kernel methods where the breakpoints were formerly set. Normally, one has to patch these methods to create callbacks forwarding required information into the high-level debugging environment. To remain kernel patch-free, we use library interposition and preload a shared library which overwrites appropriate SystemC kernel methods. There, the original implementation is extended by a callback into the debugging environment. A setting of LD_PRELOAD instructs the dynamic linker to use this library before any other when it searches for shared libraries (Fig. 8.5).

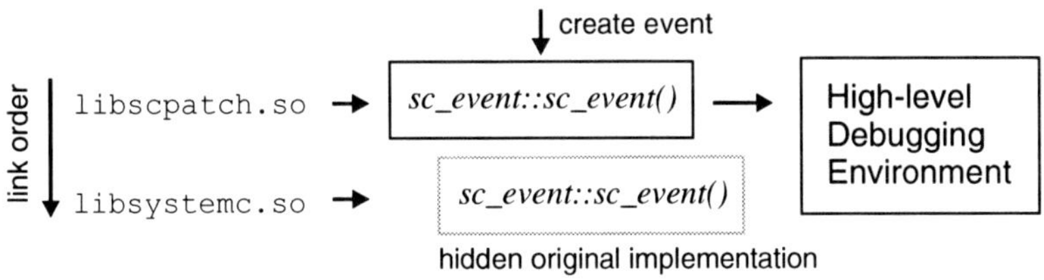

Figure 8.5. Preloading a SystemC kernel method.

Table 8.2. High-level debugging environment setup.

Test	*GDB memory usage*	*Setup time*		
		w/o	*Nonintrusive*	*Preloaded*
A[1]	45 MB	<1 s	2.25 min	1.2 s
B[2]	202 MB	~10 s	26 min	25 s
C[3]	208 MB	~10 s	> 40 min	31 s

[1]**Test system**: AMD Opteron 248 processor @2200 MHz, 3 GB real memory, **Test application**: multiple instances of a simple producer/consumer application, #threads: 204, #methods: 1010, #sc_events: 3638.
[2]**Test system**: AMD Athlon XP processor @1800 MHz, 3 GB real memory, **Test application**: bus interface controller for various protocol implementations, #threads: 29, #methods: 56, #sc_events: 306.
[3]**Test system**: see 2., **Test application**: serial interface, #threads: 31, #methods: 136, #sc_events: 701.

Preloading works only for noninlined class methods. Hence, minor transparent changes to SystemC kernel source code were necessary, i.e. moving inlined constructors of classes `sc_signal`, and `sc_event` from header to implementation files. Time measurements (Table 8.2) document the efficiency of the new approach.

4. Practical Application

Debug Problem

As a short example we try to investigate why an interface bus of our design under test (DUT) has a value contention during simulation. It seems that there are two processes concurrently driving data onto the bus.

We know of the first process in our SystemC environment driving an initialization value after reset, namely, **tb.bif.fsm._fsm_reset()**. This thread is sensitive to the negative edge of the reset signal **fsm_rst_l**, being a low active input to our DUT.

Conventional Debug Procedure

To find the second, yet unknown, process colliding with ours, we would set a breakpoint into the unique driver function that is invoked by every module that wants to stimulate the interface bus.

On any stop at this breakpoint we then had to trace back the invoking module, e.g. with the **up** command in GDB. This can turn out to be a time consuming task, potentially ending in different modules not involved in this specific issue. Since we know the signal event on which the problem occurs, a tracing of the signals SystemC event **fsm_rst_l.m_negedge_event** would help a lot.

High-level Debug Procedure

So we restart GDB with the high-level debugging environment to use the provided commands for event tracing (Table 8.1). First, we set a breakpoint onto the observed signals negative edge event:

```
(gdb) ebreak "tb.bif.fsm_rst_l.m_negedge_event"
*** scheduled break on event
*** type <continue> to set breakpoint(s)
(gdb) continue
```

which, on the breakpoint stop, gives us two thread processes sensitive to it:

```
*** event "tb.bif.fsm_rst_l.m_negedge_event"
triggered ...
breakpoint at thread 21
breakpoint at thread 16
(gdb)
```

Knowing that thread 21 is the FSM reset process, we look into the source code thread 16 is pending with **lst**:

```
(gdb) lst 16
--lst: list active source of [c]thread/method---
process tb.bif.if_bfm._update is currently
at /home/hld/project/tb/src/if_bfm.cpp:128
in if_bfm::_update
126      get_bus_val(bus_val);
127      if_bus->write(bus_val && 0xff);
128      wait();
129   }
```

In line 127 we find a concurrent writing of a wrong value onto the bus. With the **lpt** command we review the threads sensitivity list:

```
(gdb) lpt
thread process sensitivity list
--------------------------------
tb.bif.if_bfm._update
   <dynamic> tb.bif.ch_m._update.m_value_changed
   <static> tb.bif.fsm_rst_l.m_negedge_event
   <static> tb.bif.fsm_tx_w.m_posedge_event
```

We see that the sensitivity falsely includes also the reset signal, which is not desired and turns out to be an environment bug. Compared to the conventional debug procedure we needed far less debug steps and straighter tracked down the issue.

5. Conclusion and Future Work

In this paper, we presented an environment to debug SystemC applications at a high level working with signals, ports, processes, and events. High-level debugging commands realize debug patterns typically used at AMD. The special feature of our approach is its nonintrusive implementation that means it avoids real patches of the SystemC kernel or GDB sources. We apply library interposition to preload a shared library that allows to smoothly gather required debugging information provided by the SystemC kernel. Practical experiences in an industrial design flow show promising results with only a marginal increase of debug setup time.

Future work will improve the overall performance, and implement additional commands. Also, it would be necessary to increase the abstraction level of the commands in order to further simplify debugging at SystemC level. Furthermore, establishing a methodology or cookbook to apply high-level debugging commands could help the designer finding bugs more quickly.

References

[1] ARM Ltd. MaxSim Developer home. *www.arm.com*. 2006.

[2] F. Balarin, Y. Watanabe, H. Hsieh, L. Lavagno, C. Passerone, and A. Sangiovanni Vincentelli. Metropolis: an integrated electronic system design environment. *IEEE Computer*, 36(4), April 2003.

[3] K.-C. Chen. Assertion-based verification for SoC designs. *Proc. of 5th International Conference on ASIC*, vol. 1, October 2003.

[4] L. Charest, M. Reid, E.M. Aboulhamid, and G. Bois. *A Methodology for Interfacing Open Source SystemC with a Third Party Software*. DATE 01, March 2001.

[5] F. Doucet, S. Shukla, and R. Gupta. *Introspection in System-Level Language Frameworks: Meta-level vs. Integrated*. DATE 03, March 2003.

[6] GDB home. *www.gnu.org/software/gdb*. 2006.

[7] D. Große, R. Drechsler, L. Linhard, and G. Angst. *Efficient Automatic Visualization of SystemC Designs*. FDL'03, September 2003.

[8] E. Gamma, R. Helm, R. Johnson, and J. Vlissides. *Design Pattern – Elements of Reusable Object-Oriented Software*. Addison-Wesley Professional Computing Series, 1999.

[9] GRACE++ project home. *www.iss.rwthaachen.de/Projekte/grace/ visualization.html*. 2006.

[10] D.D. Gajski, J. Zhu, R. Domer, A. Gerstlauer, and S. Zhao. *SpecC: Specification Language and Methodology*. Kluwer Academic, 2000.

[11] Open SystemC Initiative home. *www.systemc.org. 2006.*

[12] CoWare ConvergenSC System Verifier home. *www.coware.com*. 2006.

[13] Synopsys System Studio home. *www.synopsys.com*. 2006.

Chapter 9

TRANSACTION-LEVEL MODELING IN COMMUNICATION ENGINE DESIGN

A Case Study

Vesa Lahtinen, Jouni Siirtola, and Tommi Mäkeläinen
Nokia Research Center

Abstract This chapter presents a case study of using transaction-level modeling (TLM) for architecture exploration and hardware/software (HW/SW) codesign in developing communication engines. It will be shown that considerable time savings are possible in analyzing architecture modification effects and providing early information from HW designers to SW designers, and vice versa. What is required is a standard and well-documented way to do the modeling. Some modifications and additions to the traditional design flows are needed, but the benefits far outweight the drawbacks. Both the HW and the SW designers found this new approach appealing and well worth the extra effort.

Keywords Transaction-level modeling, system-level design

1. Introduction

One of the most demanding tasks in developing complex embedded systems is the coordination of hardware (HW) and software (SW) efforts. In an ideal situation, these two efforts could progress in parallel without delaying each other. In addition, the early analysis of various architecture choices is a problematic task: on the one hand, the results need to be reliable, but on the other, the analysis needs to be conducted with tight time schedules.

These problems are particularly apparent in designing communication engines, which are extremely complex embedded systems with tightly coupled HW and SW parts, highly optimized structures, and a wide range of internal and

S.A. Huss (ed.), Advances in Design and Specification Languages for Embedded Systems – Selected Contributions from FDL'06, 145–156.

external HW/SW intellectual property (IP) components. The SPIRIT (Structure for Packaging, Integrating and Reusing IP within Tool-flows) standard [1] is an attempt to ease some of these issues. An additional problem is caused by the mixture and evolvement of the standards that these engines support. Support of new features needs to be added to the HW rapidly and the effect on SW needs to be visible immediately after the modifications without any low-level HW redesign time.

Transaction-level modeling (TLM)[2, 3, 4] has been proposed as a way to tackle both of the presented problematic design issues: HW/SW codesign and early architecture analysis. Two standards that are closely related to TLM are the open SystemC initiative (OSCI) [5] and the open core protocol (OCP)[6], that standardize a modeling language and an IP interface, respectively. The existence of two standards with slightly different approach to TLM has been problematic although some propositions to combine these two have emerged [7].

This Chapter presents a case study of utilizing TLM in designing communication engines. The structure of the Chapter is as follows. Section 2 describes the methodology used after which Section 3 presents the test case. The experiences of the case example are given in Section 4 after which Section 5 concludes the Chapter.

2. Transaction-Level Modeling

Background

The goal of transaction-level modeling is to speed up simulation time and the time it takes to develop the models. The TLM style advocated here is referred to as layer 2 in OCP and programmers' view with time (PV+T) in OSCI TLM [7]. After this, the term TLM is used as a synonym of combined layer 2 and PV+T style modeling. The main difference to the traditional register transfer level (RTL) model is the removal of clock and pin accuracy. Where RTL is clock cycle and pin accurate, TLM is event-based (cycle approximate with annotated delays) and models data as a collection of signals. Because of the simplifications, TLM models are faster to design than the corresponding RTL ones.

One drawback of the TLM model is that it cannot be used as an input to a traditional synthesis flow. This implies that two separate models of the system components are required: RTL model for implementation and TLM for HW/SW codesign and early architecture analysis. The benefits and drawbacks of this approach are discussed in Section 4.

It should also be noted that there is nothing unique or outstanding in the ability of SystemC to model inherently concurrent HW systems. In fact, the debugging and timing behavior have some apparent limitations. The crucial thing here is, however, that SystemC has evolved into a standard with a developing community and few agreed-upon styles in which to model.

Separation of concerns has been the key thesis of many platform-based design approaches. Usually this refers to the separation of communication from computation but it can also be used to refer to the separation of the physical and the logical architecture. This is an important aspect of architecture specification in a platform development process.

Related Work

Many research groups have presented their own viewpoints on the basic principles of SystemC usage. The use of SystemC for cosimulation and emulation has been documented by Benini [2]. In addition, an overview of TLM is presented by Cai [3] and an overview of TLM flows by Donlin [4]. Different approaches to SystemC modeling are presented by Colgan [7] and Kogel [8]. The differences between RTL and TLM have been studied by Calazans [9].

SystemC-based transaction level modeling has been widely adopted in the recent years. Many of the presented use cases have concentrated on one aspect of the whole system. These include the scheduling [10] and modeling [11] of operating system, bus architecture modeling [12], communication architecture exploration [13], and multiprocessor system exploration [14].

The capability of SystemC to help in capturing the behavior of complex systems has been demonstrated with many test cases. In the communication device field, these test cases include a WLAN (Wireless Local Area Network) system [15] and an UMTS (Universal Mobile Telecommunication Services) modem [16]. In addition, Xu [17] presents a system-level TLM example for mixed language simulation.

3. Case Study: NeMo Project

Project Background

The new modem (NeMo) project was an experiment in the presented system modeling issues. It was conducted during the years 2002–2005 in the Nokia Research Center. The goal of the project was to develop a new platform architecture for communication engines and to model it with SystemC. The primary goal was a WCDMA (Wideband Code Division Multiple Access) modem but also multimode issues were taken into account in the architecture design.

The targets for the architecture development experiment were:

- Support for early SW development
- Support the modular composition of an architecture
- Separate logical and physical architecture development
- Early integration and verification of the platform

- Coordinated management of design objectives and requirements
- Maintenance of multiple abstraction levels for architecture models
- Support for architectural exploration and performance evaluation

Of the targets the first was seen as the most important one. The use of modeling for architecture exploration and performance evaluation was given a smaller weight. The use of multiple abstraction levels was a practical requirement due to the various models offered by the IP vendors.

The NeMo Architecture

Figure 9.1 depicts a simplified block diagram of the actual NeMo platform architecture. Modularity was seen as a key point, because of the requirement to support multiple wireless standards. It was emphasized in the architecture with separated clusters for all standards. In addition, the use of shared memory for all the clusters was studied to make the memory utilization more efficient compared to using dedicated memories for each cluster. Two controllers handled the high-level control of the platform: one was used for resource management and the other for controlling the data communication in the system. In addition to the shared memory and the controllers, also external interfaces to DRAM/FLASH, application engine (APE), and RF platform are common for all the clusters.

The modularity was also emphasized in the example design of the WCDMA cluster. It had four receiver (Rx) and two transmitter (Tx) side processing blocks and one internal block for the RF connection. The Rx blocks were used for medium access control (UMAC), transport channel decoding for normal (TrCH) and high-speed channels (HS-TrCH), and symbol rate processing (SR). The Tx blocks, on the other hand, were used for medium access processing (UMAC) and encoding and symbol rate processing (TrCh). All the blocks had a similar internal structure with an algorithm data path, a control unit, and a common communication scheme using a component-internal communication block, the OCP-based interconnect, and the shared memory. Particularly OCP as a standardized communication interface was a key enabler for the modular design of the architecture.

The blocks shown with gray shading in Fig. 9.1, were external IP blocks. SystemC models of them were also provided by the vendors, but they used a more accurate (clock-cycle accurate) timing than was used for the internal modeling. This required special transactor blocks to the crossover points which, unfortunately, slowed down the simulations.

The Utilized Modeling Process

The modeling process utilized in the NeMo project is depicted in two phases in Figs. 9.2 and 9.3. The basic idea was to change from "bottom-oriented" to "top-oriented" design flow. This implies that some extra work is required in the

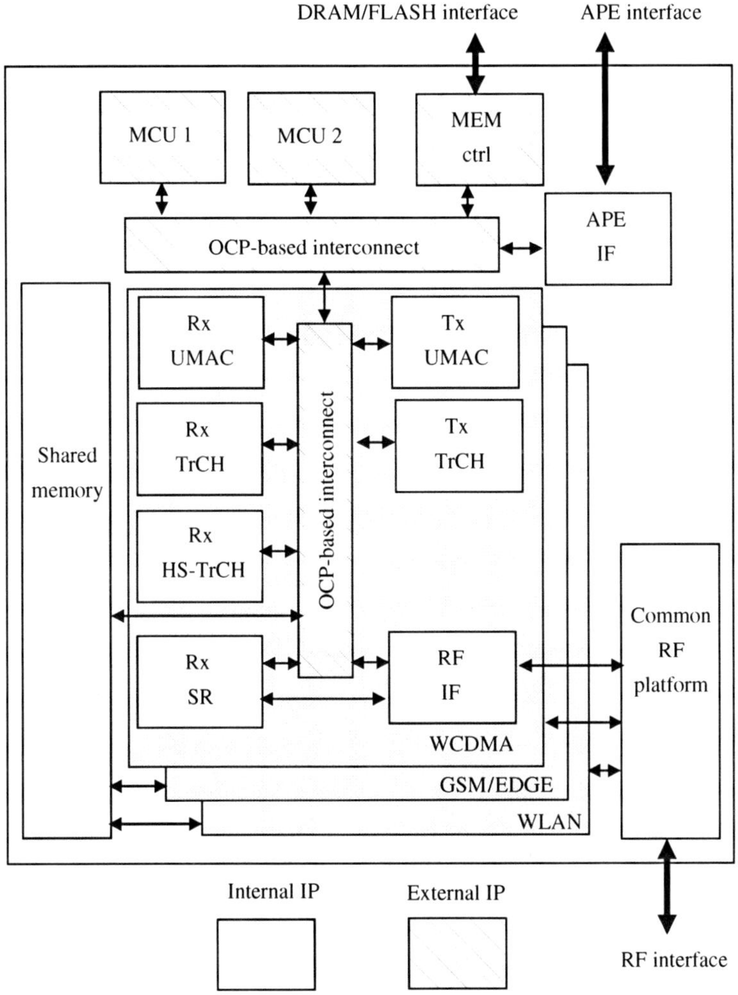

Figure 9.1. The architecture developed in the NeMo project.

start, but also, that it should result in easier integration and testing in the final stages of the design process. Another characteristic is that the design process is not only seen as HW and SW development. A new phase, entitled architecture development precedes and steers them both.

The platform architecture is developed mainly in the cospecification phase. The inputs to this phase include objectives and requirements of the project. They are used to formulate the operational concept and to find out what kind of algorithms and protocols are required. After this, the main phase of the platform architecture development, namely cospecification can start.

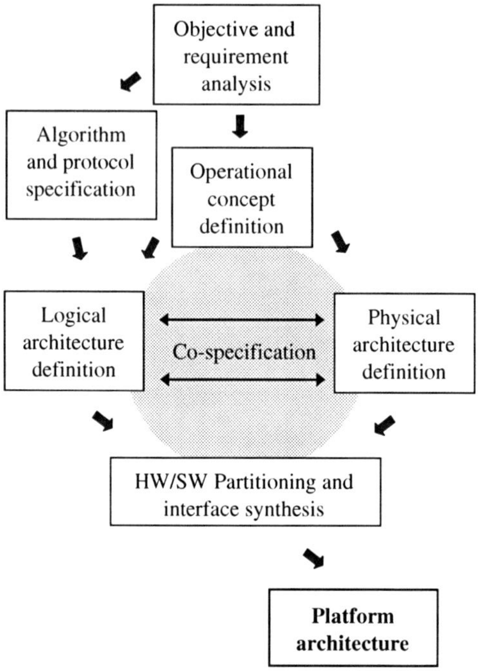

Figure 9.2. NeMo design process: Cospecification.

In the cospecification phase, the logical architecture, and the physical architecture development progress in parallel. The logical architecture defines the functionality and the functional clustering used; in this case using pure C++. The physical architecture, on the other hand, defines the physical resources, such as processing units, storage units, and interconnects, of the platform. The result is the platform architecture, which contains the HW/SW partitioning of the logical architecture and the mapping of it to the physical architecture. The SW parts remain as C++ and the HW is described in (TLM) SystemC.

The main task of the cospecification phase is to conduct early architecture analysis. This requires that modifications to the architecture are easy and, therefore, extensive HW/SW mapping and component allocation experiments can be conducted. The accuracy of the results is, however, not a top priority at this stage.

The platform architecture is represented as a virtual platform, which acts as an executable specification to the HW development process, and as an environment for SW development. Therefore, it forms the basis for the codesign phase, which produces the actual communication engine architecture implementation using the standard HW and SW implementation flows.

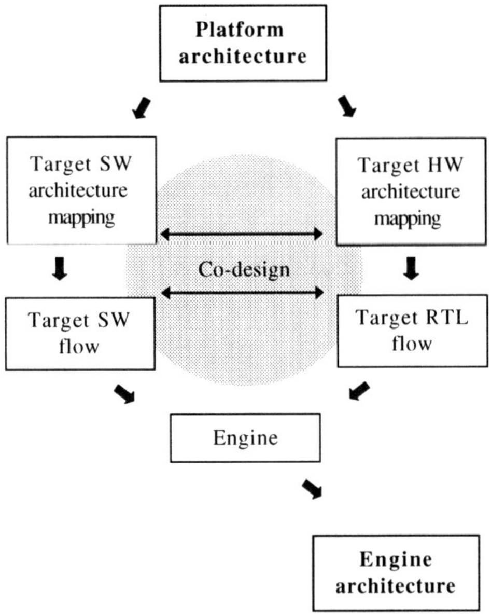

Figure 9.3. NeMo design process: Codesign.

The main requirement to the virtual platform is the functional accuracy from the SW point of view. This means that the utilized modem blocks could be modeled with only their register interface and the internal algorithm behavior. Timing accuracy is based on delay estimates, rather than clock-cycle accuracy. This implies that short run-times of sufficiently extensive simulations (as required by e.g. the porting of operating systems) is preferred over accurate modeling of time.

4. Experiences of the Study

Feedback From Designers

The feedback from HW and SW design teams was very positive. The virtual platform for SW designers gave them a development environment before the RTL was ready for RTL simulations and emulations. On the other hand, the early feedback from SW designers was perceived to be extremely important by the HW designers. Getting feedback only at the emulation phase had previously been seen as a serious problem.

The benefits seen in the approach were:

- SW development can start early. This enables early feedback to HW designers before they have reached the prototyping phase. This makes it possible to do faster design iterations.

- The model acts as an executable HW specification and testbench and enables early, although rough, performance estimates. This minimizes the number of required iterations.
- It makes derivative product creation very fast. In these cases the models are inherited from previous platform design projects and only require a small amount of redesign.
- The model enables fast simulation times compared to RTL simulations or even emulation using dedicated HW.

The perceived drawbacks of the approach were:

- The approach is too slow for analyzing radical changes in the architecture. The initial work to build the required models takes a long time and modifications are not always straightforward.
- Maintaining and writing the models causes extra work to the traditional design flows. In order to make the model creation and simulation time fast they have to be made separately from HW design. The responsibility for the model and implementation equivalence was also seen problematic.
- The biggest obstacle was the addition of new phases to the traditional design flow which means new tools, new methodologies, new licenses, and so on. A new mind-set and a lot of training are required for a successful adoption of these new design techniques and their integration into the current design flows.

In addition, some practical issues were raised during the development process. In order for the modeling to be possible, tight control over the used modeling style is required. Internally, it is possible to define a common modeling style and model interface, but the use of external models leads, in practice, to the use of different abstraction levels and definitions. Although adapters exist to get around this, it causes extra work, problems, and slows down the simulations. For some of the IP blocks, it proved to be very difficult to acquire the required models.

The standardization of interfaces is a key requirement for successful platform design. However, it can be troublesome if all the IP providers do not support the same set of standards and features. This requires protocol adapters and strict internal specifications on preferred interface standard and the features that can and should be used. In addition, some IP providers offered cycle accurate models that did not behave in the same manner as their RTL implementation. The verification of the equivalence between the IP model and the actual implementation is one thing that needs to be developed further.

The synthesis of RTL descriptions from TLM SystemC is sometimes seen as a required feature for the whole approach to be accepted. Several commercial tool sets have emerged for this purpose. Although it is unclear whether this can happen in a system scale, at least the synthesis of algorithm blocks from their C-based models has been demonstrated to work. At least some form of synthesis would be beneficial in order to gain the support of traditional HW flow supporters for TLM.

There are also some issues in the available toolsets. They are still quite immature, although developing all the time. A problematic issue is that the tools have some proprietary solutions for features that are not covered by standards yet. This is an understandable way in the development phase of these methods but unacceptable when they are to be brought into production use. Also the licensing can be expensive, since the goal is to use these tools for SW development with potentially a large number of users.

To introduce TLM into a design process, a standardized and well documented modeling style is needed. This includes specification of the used coding structures, abstraction levels, timing models, and interfaces. These kind of modeling guidelines are well developed in the RTL space but still require effort in TLM, particularly since the chosen language is C++-based SystemC which offers numerous ways to model in an undebuggable and cryptic way.

Benefits of the New Design Style

Figure 9.4 depicts what can be achieved with the SystemC-based TLM modeling. The upper part of the figure represents the traditional approach where functional specification of the system is followed by HW design. SW design can only start when RTL of the HW is ready. The HW/SW integration and verification is done in RTL level and emulation. This is quite slow and if problems arise they are quite troublesome to solve. In addition, some errors found in this last stage can lead to modifications of the functional design. All in all, this approach can lead to unacceptably long design iteration times.

The first modification to the original flow is to start the design process with architecture specification. This can prolong the functional design phase, but the SW development can start months before in the traditional approach due to shorter HW development times. In addition, early architecture exploration should diminish the number of functional modifications.

The last modification introduces fast processor models. These enable the parallel advancing of HW design, SW development, and HW/SW integration and verification. The most crucial thing that this modification offers is the ability to do early SW development before the RTL is ready. The use of fast processor models makes it possible to run complex SW simulations in acceptable speeds, i.e. close to real-time. The goal of this approach is to cut the design time of a communication engine in half.

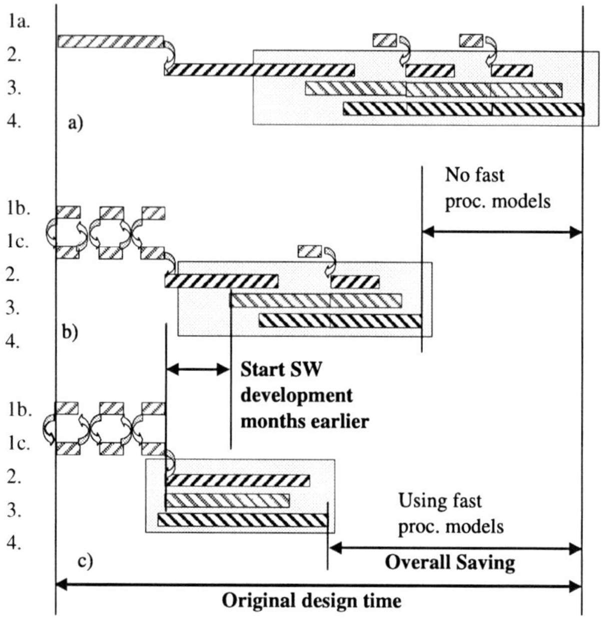

1a. Functional specification
1b. Logical architecture specification
1c. Physical architecture exploration
2. HW design
3. SW development
4. HW/SW integration & verification

Figure 9.4. Benefits of logical architecture and architecture exploration.

5. Conclusions

The problems of traditional design flows that rely on RTL for cosimulation and emulation in any large-scale design projects are evident. Too much time is spent in waiting for the RTL to finish and iterations are too costly. One solution to these issues is the use of transaction level modeling and SystemC.

Separation of logical and physical architecture enables an efficient architecture specification process that is the starting point of a TLM based modeling flow. This specification can then be used as a platform for SW development and an executable specification and testbench for HW design. Because the SW development and HW/SW integration and verification can start much earlier, significant amount of time is saved although the initial architecture design requires extra work.

To introduce TLM into a design process, a standardized and well-documented modeling style is needed. This includes specification of the used coding structures, abstraction levels, timing models, and interfaces. With these guides, both the HW and SW designers found it easy to adapt this new approach, and they both saw the extra work it requires justified.

Acknowledgments

In addition to the authors of this chapter, the following researchers were major contributors to the NeMo project: Adama Bamba, Dirk Bierbaum, Arto Hännikäinen, Pasi Katajainen, Anssi Marttila, Jukka Nurminen, Jaakko Varteva, Ville Pernu, Jaakko Tölli, Tiina Yli-Pietilä, and Timo Yli-Pietilä.

References

[1] Spirit Consortium, Spirit-User Guide v1.2, (November 6, 2006); www.spiritconsortium.com.

[2] L. Benini, D. Bertozzi, D. Bruni, N. Drago, F. Fummi, and M. Poncino, SystemC cosimulation and emulation of multiprocessor SoC design, *IEEE Computer*, April 2003, pp. 53–59.

[3] L. Cai and D. Gajski, Transaction level modeling: an overview, *Proc. Intl. Conf. Hardware/Software Codesign and System Synthesis*, October 2003, pp. 19–24.

[4] A. Donlin, Transaction level modeling: flows and use models, *Proc. Intl. Conf. Hardware/Software Codesign and System Synthesis*, September 2004, pp. 75–80.

[5] Open SystemC Initiative OSCI, SystemC Documentation (November 6, 2006); www.systemc.org.

[6] OCP-IP, Open Core Protocol, OCP 2.1 Specification, (November 6, 2006); www.ocpip.org.

[7] J. Colgan and P. Hardee, *Advancing Transaction Level Modeling – Linking the OSCI and OCP-IP Worlds at Transaction Level*, Open-Systems Publishing (November 6, 2006); www.opensystems- publishing.com/whitepapers.

[8] T. Kogel, A. Haverinen, and J. Aldis, OCP TLM for Architectural Modeling. Methodology: White Paper, OCP-IP (November 6, 2006); www.ocpip.org.

[9] N. Calazans, E. Moreno, F. Hessel, V. Rosa, F. Moraes, and E. Carara, From VHDL register transfer level to SystemC transaction level modeling: a comprehensive case study, *Proc. Symp. Integrated Circuits and Systems*, September 2003, pp. 355–360.

[10] H. Yu, A. Gerstlauer, and D. Gajski, RTOS Scheduling in Transaction Level Models, *Proc. Intl. Conf. Hardware/Software Codesign and System Synthesis*, October 2003, pp. 31–36.

[11] H. Posada, J. Adamez, P. Sanchez, and E. Villar, POSIX modeling in SystemC, *Proc. Asia and South Pacific Conf. Design Automation*, January 2006, pp. 485–490.

[12] M. Caldari, M. Conti, M. Coppola, S. Curaba, L. Pieralisi, and C. Turchetti, Transaction-level models for AMBA bus architecture using SystemC 2.0, *Proc. Design Automation and Test in Europe*, March 2003, pp. 26–31.

[13] S. Pasricha, N. Dutt, and M. Ben-Romdhane, Fast exploration of bus-based On-Chip communication architectures, *Proc. Intl. Conf. Hardware/ Software Codesign and System Synthesis*, September 2004, pp. 242–247.

[14] A. Wieferink, M. Doerper, R. Leupers, G. Ascheid, H. Meyr, T. Kogel, G. Braun, and A. Nohl, System level processor/communication co-exploration methodology for multiprocessor system-on-chip platforms, *IEE Proc.-Comput. Digit. Tech.*, January 2005, pp. 3–11.

[15] J. Lee and S.-C. Park, Designing SoC communication architectures for WLAN system using SystemC, *Proc. Intl. Symp. Communications and Information Technologies*, October 2004, pp. 536–540.

[16] P. Martinelli, A. Wellig, and J. Zory, Transaction-level prototyping of a UMTS outer-modem for System-on-Chip validation and architecture exploration, *Proc. Intl. Works. Rapid System Prototyping*, June 2004, pp. 193–200.

[17] S. Xu and H. Pollit-Smith, A TLM platform for System-on-Chip simulation and verification, *Proc. Intl. Symp. VLSI Design, Automation and Test*, April 2005, pp. 220–221.

Chapter 10

OBJECT-ORIENTED TRANSACTION-LEVEL MODELLING

Martin Radetzki
Institut für Technische Informatik,
Pfaffenwaldring 47,
70569 Stuttgart, Germany
martin.radetzki@informatik.uni-stuttgart.de

Abstract This chapter presents a contribution to the methodology of transaction-level modelling (TLM) based on SystemC and the TLM standard of the Open SystemC Initiative (OSCI). Different from previously published approaches to the use of TLM in conjunction with SystemC, we employ object-oriented features to represent transactions, namely an inheritance relationship of transaction classes, transaction polymorphism, and dynamic binding of transactions to methods of SystemC modules. Advantages of this new modelling style include reduced programming effort for transaction dispatch, easier extensibility, and guaranteed consistency. The extension of a simple transaction model to cover burst transactions is demonstrated. We make use of concepts adapted from the concurrent object-oriented design pattern known as active object. Our approach allows to fully utilize SystemC / TLM features and is in no conflict with the standardized aspects of this methodology.

Keywords Transaction-level modelling, SystemC, object-oriented, embedded systems

1. Introduction

The complexity of developing up-to-date embedded systems makes it necessary to model these systems prior to their implementation. This helps to ensure functional correctness and to estimate performance requirements posed upon the implementation platform by the application before starting the costly and time-consuming implementation process. Besides system-level languages such

S.A. Huss (ed.), Advances in Design and Specification Languages for Embedded Systems – Selected Contributions from FDL'06, 157–174.

as SpecC [2] and SystemVerilog [6], the C++ class library SystemC [5] together with the Transaction-Level Modelling (TLM) library [11] provide a basis for these modelling activities.

SystemC allows the user to specify modules that encapsulate functionality. An instance of a module is an object in the C++ sense. SystemC mechanisms are available to specify that a module has one or more threads. An instance of a module with at least one thread is called active object in this context, meaning that the object has computational resources of its own. Moreover, SystemC provides communication mechanisms such as ports and channels. These can be utilized to model classic signal communication as in hardware description languages (HDL) or to let objects communicate via the so-called interface method call (IMC) paradigm. Using the latter, the user is free to derive any desired communication interfaces from a SystemC interface class.

The TLM methodology by the Open SystemC Initiative (OSCI) intentionally restricts this freedom, aiming to give users more guidance and to achieve better interoperability of models. For this purpose, the SystemC TLM library provides standardized communication interfaces for transaction-level modelling of digital systems.

Transaction-level models consist of SystemC modules that initiate transactions (masters, initiators) and modules that receive and may respond to transactions (slaves, targets). A network that connects the initiators and targets enables their communication. This network may more or less closely resemble bus architecture and timing behaviour of the implementation platform, depending on the purpose of modelling:

- Functional view (FV) models employ untimed point-to-point communication, focusing purely on functional behaviour.

- Programmer's View (PV) models use blocking transactions and passive targets to provide a basis for early software development. Timing may be modelled coarsely, bus arbitration is typically not modelled.

- Architecture View (AV) – also called cycle-approximate (CX) – models resemble the bus architecture and arbitration of an implementation platform with approximated timing. They are used for timing / performance estimation.

- Verification View (VV) – also called cycle-accurate (CA) – models are clocked and represent the exact bus behaviour in each cycle. Such models serve as a reference against which synthesizable RTL implementations are verified.

In AV and VV, typically non-blocking transactions are modelled, meaning that the initiator can resume its operation while a transaction is transported to and processed by a target module. This makes it necessary for the target to have computational resources of its own, i.e. to be modelled as an active object. Such an object must receive transaction messages from the bus system and invoke the corresponding internal functionality. In the following, we present a new, advantageous way of modelling the transactions, their initiation, and their dispatch in SystemC.

An overview of related work is given in the Section 2. In Section 3, our solution is presented with a focus on modelling of simple transactions and its differentiation from the state-of-the-art TLM. Section 4 shows an extension to cover non-blocking burst transactions in the model. Section 5 presents experimental results, in particular on simulation performance in relation to other approaches, and Section 6 concludes the presentation of our work.

2. Related Work

The Active Object Design Pattern

The concurrent object-oriented design pattern known as active object is described, among other sources, in [15]. It enables concurrent objects (clients) to request the execution of services by a target object (server) that has a control flow (thread) of its own. To this end, the following roles are defined as part of the design pattern and have to be implemented by respective classes:

- *Servant*: implements the functionality and the thread of the server. This corresponds to the role of a target (bus slave) in a transaction-level model while the client is an initiator (bus master).

- *Proxy*: provides to the clients an interface for requesting services. This role is assumed by the bus in our modelling approach.

- *Method request*: message sent by the proxy to the scheduler in order to initiate the execution of a service, corresponding to a bus transaction.

- *Scheduler*: determines which service is executed by the servant at which point in time and arbitrates concurrent requests. In the bus model, the arbitration scheme of the bus assumes this responsibility.

- *Future*: an object provided by the proxy to the client in order to enable the client to obtain the results of a requested service at a later point in time, when these results are available. This enables the client to proceed after requesting a service without having to wait for the service results. In the bus model, the future concept is implemented as part of a transaction.

Object-Oriented System Modelling

In the context of approaches towards the object-oriented extension of HDL, in particular of VHDL, concepts from concurrent object-oriented programming have been considered and adapted to the field of hardware modelling and design [17].

A mechanism for initiation and dispatch of service requests has been described in [4] as part of a SystemC extension on which the ODETTE System Synthesis Subset (OSSS) is based. This mechanism is built into a class/macro library so that the user does not have to write or modify dispatch code. Thereby, extensibility issues are addressed. Moreover, message dispatch to polymorphic objects – objects whose exact class type is not known at the time of compilation – is provided as a feature that supports flexible and extensible models. With proprietary tools, the message exchange can even be synthesized. However, the OSSS library is neither an OSCI standard nor compatible with the TLM methodology and library. It covers modelling of communication on a level comparable to the programmer's view (PV) but is not suitable to model actual bus structures.

Active objects have been described as a feature of Objective VHDL [8, 13], an object-oriented extension to VHDL. Message passing to active objects is not built into that language, but can be modelled in an object-oriented fashion as described in [12]. This approach features the idea of using object-oriented mechanisms to model messages and their dispatch. It has inspired the solution presented in the remainder of this chapter. New aspects of our contribution include the adaptation to C++ and SystemC, use of the TLM library in accordance with TLM methodology, a systematic path from passive objects (PV models) to active objects (AV models), and the modelling of non-blocking transaction communication.

Transaction-Level Modelling

Transaction-level modelling (TLM [3]) is a methodology independent of a specific language. It is supported by system description languages such as SystemC [10], SpecC [2] and SystemVerilog [6]. In this contribution, we use SystemC in conjunction with the SystemC TLM library [11] (in the current version 1.0) to implement our approach towards object-oriented transaction-level modelling.

Bus protocols, e.g. the AMBA AHB protocol by ARM [1] or the Open Core Protocol (OCP [9]) have been described with SystemC. An investigation of the resulting models shows that, while their interfaces are described on transaction level, the underlying implementation often is on a significantly lower level that hardly yields an abstraction from the signal/RT level. In this case, high simulation performance is not achieved by abstraction but by an implementation precise, it is proportional to the number of atomic properties). The case study implied many observers (i.e. LTL formulae to check) verifying a fairly simple model. Of course, verifying the same properties on a more complex model would lower the impact of the observers overhead. Furthermore, the size of each formula tends not to exceed a certain size in practice. On the other hand, to evaluate advantages of doing introspection during the initialization phase (i.e. static introspection), the simulation was also performed using dynamic introspection (i.e. introspection during the evaluation of properties). In this case, the overhead rate due to introspection is increased to 90%. Concerning the model, the verification allowed us to find limit cases and synchronization bugs in our implementation. The hardest and longest part was not the construction of the observers but the construction of the model itself. Correction of bugs remains especially difficult. This case study was also a good opportunity to pinpoint weaknesses of the verification engine and find solutions to improve its performances. Among others, the formulation of complex behaviors with LTL is a difficulty, frequently encountered in formal verification. However, the use of formula patterns [15] alleviates this task. Our LTL formula editor was thus extended to propose a set of such patterns to assist the designer.

5. Conclusion

This work presents a verification method and tools with their implementation in a system-level design environment. Our approach brought to light the steps of the verification process which could benefit from introspection and thus greatly simplify the development of modeling and simulation environments. Introspection is used to get the structure of the model and to retrieve data during simulation. ESys.NET can easily be extended thanks to hook-points to synchronize the execution of observers with model simulation.

References

[1] it SystemC Version 2.1, http://www.systemc.org/, 2006.

[2] Rich, D. I., The evolution of SystemVerilog, *Design & Test of Computers*, IEEE , 20(4) pp. 82–84, July–August 2003.

[3] Lapalme, J., Aboulhamid, E. M., and Nicolescu, G. (2006). A new efficient EDA tool design methodology, *Trans. on Embedded Computing Sys,* 5, 2 (May. 2006), 408–430.

[4] it .NET Framework, http://www.microsoft.com/, 2003.

[5] Doucet F., Shukla S., and Gupta, R. (2003). Introspection in system-level language frameworks: meta-level vs. integrated. In: *Design, Automation and Test in Europe Conference and Exhibition*, pp. 382–387.

[6] Patel, H. D., Mathaikutty, D. A., Berner, D., and Shukla, S. K. (2005). SystemCXML: An Extensible SystemC Front End Using XML, Technical Report No: 2005-06. http://systemcxml.sourceforge.net/

[7] Moy, M., Maraninchi, F., and Maillet-Contoz, L. (2005). PINAPA: an extraction tool for SystemC descriptions of systems-on-a-chip. In: *Proc. the 5th ACM international Conference on Embedded Software.*

[8] Norris Ip, C., and Swan, S. (2005). *A Tutorial Introduction on the New SystemC Verification Standard,* www.systemc.org.

[9] IEEE P1850 standard for PSL – Property Specification Language, http://www.eda.org/ieee-1850/

[10] Pnueli, A. (1977), The temporal logic of programs. In: *Proc. the 18th IEEE Symposium on Foundations of Computer Science,* pp. 46–67.

[11] Wolper, Pierre, Vardi, Moshe Y., and Sistla, A. Prasad (1983). Reasoning about infinite computation paths. In: *Proc. 24th IEEE Symposium on Foundations of Computer Science,* pp. 185–194.

[12] Giannakopoulou, D. and Havelund, K. (2001). Automata-based verification of temporal properties on running programs. In: *Proc. the 16th IEEE International Conference on Automated Software Engineering,* Automated Software Engineering, IEEE Computer Society, Washington, DC, 412.

[13] C. Eisner, D. Fisman, J. Havlicek, Y. Lustig, A. McIsaac, and D. Van Campenhout. Reasoning with temporal logic on truncated paths. In: *Proc. International Conference on Computer Aided Verification (CAV),* LNCS 2725, pp. 27–39.

[14] Gastin, Paul, and Oddoux, Denis (2001). Fast LTL to Büchi Automata translation. In: *Proc. International Conference on Computer Aided Verification (CAV),* Lecture Notes in Computer Science 2102, p. 53–65.

[15] Dwyer, M. B., Avrunin, G. S., and Corbett, J. C. (1999). Patterns in property specifications for finite-state verification. In: *Proc. the 21st International Conference on Software Engineering,* (Los Angeles, California, United States, May 16 – 22, 1999), EEE Computer Society Press, Los Alamitos, CA, p. 411–420.

[16] ARM Corporation. http://www.arm.com/

Chapter 14

FORMALIZING TLM WITH COMMUNICATING STATE MACHINES

Bernhard Niemann[1], Christian Haubelt[2], Maite Uribe Oyanguren[1], and Jürgen Teich[2]

[1] *Fraunhofer Institute for Integrated Circuits*
Am Wolfsmantel 33
91058 Erlangen, Germany

[2] *University of Erlangen-Nuremberg*
Am Weichselgarten 3
91058 Erlangen, Germany

Abstract Transaction Level Models are widely being used as high-level reference models during embedded systems development. High simulation speed and great modeling flexibility are the main reasons for the success of TLMs. While modeling flexibility is desirable for the TLM designer, it generates problems during analysis and verification of the model. In this paper we formalize the notion of Transaction Level Models by introducing a set of rules that allow the transformation of TLMs to a set of communicating state machines. SystemC being the most popular TLM language, we additionally present a finite state model of the SystemC scheduler. Finally, we demonstrate that using our modeling approach, a standard model checker can be employed to formally prove properties on the finite state model.

1. Introduction and Motivation

Transaction Level Modeling [13] using SystemC [7] has become a standard way of implementing high-level reference models for embedded systems as it allows to implement efficient and flexible virtual prototypes. Over the past years, research activities were mainly focused on exploiting modeling flexibility and exploring different levels of communication and behavior abstraction

S.A. Huss (ed.), Advances in Design and Specification Languages for Embedded Systems – Selected Contributions from FDL'06, 225–242.

(see, e.g. [1]). More recent work concentrates on formalization and verification. The aim of our work is to introduce enough formalism to Transaction Level Modeling to allow the application of formal methods like model checking but to retain enough flexibility to be able to apply our methodology to a wide range of different models.

While existing work is mostly focused on the translation of C++ functionality and SystemC constructs, we aim at a strong emphasis on communication and transactions. To this end, we translate a SystemC TLM to a set of communicating state machines. The automata proceed synchronously, meaning that a transition of the overall system consists of the simultaneous transition of all automata in the system. The state of a module in our approach is composed of *behavior state*, *initiator state*, *target state*, and *object state*. SystemC provides *events*, *time-outs*, and a *scheduler* to enable modeling of a wide variety of communication styles. Therefore, in addition to the modules itself, we provide rules for a finite state representation of scheduler, timed, and untimed events.

The result of the transformations presented in this work is a finite state model which is functionally equivalent to the original TLM and can be used as the input to a standard model checker. Contrary to other approaches, our model explicitly separates behavior from communication and transaction initiation from the target of a transaction. The advantage is a concise and standardized way of formulating properties reasoning about transactions and transaction sequences.

In the remaining part of the paper, we survey related approaches in Section 2 and formally define the terms Transaction Level Model and finite state machine as used in this work in Section 3. Section 4 is dedicated to the transformation of modules, while in Section 5, the rules for the transformation of scheduler and events are presented. Experimental results are discussed in Section 6.

2. Related Work

Related approaches are split into two closely linked fields both being subject of current research. Firstly, as we enable the formulation of properties for Transaction Level Models, work carried out in the area of assertions for TLMs has to be considered. Secondly, existing methods to apply formal verification to SystemC designs are discussed.

Regarding assertions, Ruf et al. [14] convert FLTL properties to C++ which can directly be integrated into a SystemC design. Dahan et al. [3] convert PSL properties to deterministic finite automata which are converted to VHDL or SystemC. Habibi et al. [6] propose static program analysis and genetic algorithms to increase coverage of assertions specified in PSL. All these approaches require a system clock. In [11] the application of assertions to timed and untimed TLMs is proposed. Every transaction is associated with a Boolean signal, so that abstract system properties can be specified using temporal assertions.

In the area of formal verification, Große et al. [4] apply bounded model checking to SystemC designs at the Register Transfer Level (RTL). Habibi and Tahar [5] convert an AsmL model of the TLM to SystemC. A set of properties specified in PSL is converted via AsmL to C#. At the AsmL level the properties are encoded in every state of the design and can be checked on the fly during the finite state machine generation. Peranandam et al. [12] use symbolic simulation of a Message Sequence Model to verify LTL properties and obtain coverage information. Their approach does not consider event notification and event sensitivity. Karlsson et al. [8] use a Petri net-based approach. They translate the SystemC design to PRES+ which can be used for CTL model checking after a conversion to hybrid automata. Even though translation of TLMs is possible using this approach, there is no mechanism to easily identify transactions. The focus is modeling of behavior and not communication.

Two approaches using, similarly to our approach, a set of communicating parallel automata can be found in Kroening and Sharygina [9] and Moy et al. [10]. In [9], Labeled Kripke Structures are used to represent a SystemC design. The work is focused on the application of abstraction techniques to the behavior. They do not take into account communication through transactions and restrict their discussion to clocked models communicating over signals. Translation of SystemC to Heterogeneous Parallel Input/Output Machines (HPIOM) is presented in [10]. While their model of the scheduler is similar to ours, the intention of the model and the treatment of transactions differ. They propose to use translation patterns for different TLM channels, and aim at proving the validity of safety properties reasoning about implementation behavior. We suggest a pattern for the transfer of control from the initiator to the target and treat transaction functionality like any other C++ code. Our aim is to provide a formalism to reason about transactions using temporal logics with the transactions itself being atomic propositions used within the properties. This makes our model well suited for the formulation of temporal properties reasoning about transactions.

Our methodology differentiates from the related work by presenting a model for TLMs, where an easy identification of initiator, target, and transaction is possible. While previous methodologies have focused on behavior, we make a strong emphasis on communication, and preserve the separation from communication and behavior of the SystemC TLM within our formal model. Moreover, we can handle overlapping executions of the same transaction and model arbitrary primitive channels.

3. Prerequisites

Transaction Level Models are characterized by communication through *interface method calls*, i.e. calling a method implemented in one module (the target) from within another module (the initiator). Moreover, their *level of abstraction* is above RTL, even though the communication, or parts of it, may be cycle-accurate. Formally, in this work, a TLM is a six-tuple $S := (M, M_I, M_T, N, T, I)$, where M is a set of *modules*, $M_I \subset M$ the set of *initiator modules* and $M_T \subset M$ the set of *target modules*. A module need not necessarily belong to only one category, however, $M_I \cup M_T = M$. N is the set of all *interface method names*. The set of transactions is described as $T \subseteq M_T \times N$ and associates target modules and interface method names. Note that the same interface method may be implemented differently in different target modules. Thus it is necessary to associate the name of the target module with the name of the interface method to uniquely identify a transaction. The function $I : T \mapsto 2^{M_I}$ maps each transaction to a set of initiators.

A finite automaton. is a six-tuple $A := (\Sigma, \Omega, S, \delta, \lambda, s^0)$. The *input alphabet* is given by Σ, the *output alphabet* by Ω. The *transition relation* $\delta \subseteq S \times \Sigma \times S$ describes the evolution of the state with the inputs, the *output relation* $\lambda \subseteq S \times \Sigma \times \Omega$ maps a letter of the output alphabet to each state transition. The set of initial states is given by $s^0 \subseteq S$. If the transition relation and output relation are replaced by the functions $\delta : S \times \Sigma \mapsto S$ and $\lambda : S \times \Sigma \mapsto \Omega$ and the automaton only has one initial state $s^0 \in S$, the automaton is called *deterministic*. We also use the term finite state machine (FSM) for a finite automaton.

Nondeterminism. As the execution of a SystemC TLM exhibits deterministic behavior, most of the automata in our model are deterministic. There are, however, two important exceptions. One exception is the automaton used for process activation in our model of the SystemC simulation kernel (see Section 5). Even though the *SystemC Language Reference Manual* [7] states that process activation has to be deterministic, the actual order of activation is left open to the implementation of the simulation kernel. To prove a property for all possible implementations of the SystemC simulation kernel and therefore for all legal executions of the TLM, process activation is performed non-deterministically in our model. The other exception is to use nondeterminism to replace data or address dependent conditional statements. This is an optional modification of the deterministic model, reducing the number of reachable states for the price of introducing new executions not present in the original model (see Section 6).

4. A Finite State Model for TLMs

Using the FSM model presented in this section and the model of the SystemC scheduler from section 5, it is in principle possible to describe non-TLM SystemC models. This, however, is not the intention of our approach; it is much more a consequence of the need to support multiple types of TLMs. Through its explicit modeling of initiator, target, and communication channel, it is well suited to describe TLMs but will not be the optimum choice for other modeling styles. When formally modeling, e.g. a system at RT-level, there will be no need for a scheduler or explicit modeling of the `sc_signal<>` communication channel.

A module of a Transaction Level Model is built from four basic types of FSMs: behavior, initiator, target, and object (see Fig. 14.1). Processes are described with a Behavior FSM, A_b. An initiator FSM, A_i, is needed for each process initiating a transaction, i.e. each initiator module has at least one initiator FSM. On the other hand, each initiator FSM needs a corresponding target FSM, A_t, which resides in the respective target module. Initiator and target FSM model the communication through transactions. Using one initiator-target pair per process is important to allow multiple processes to initiate overlapping executions of the same transactions. This is, to our knowledge, not currently handled by other methodologies. The object FSM, A_o, finally represents the internal state of a module and keeps track thereof. The state space of A_o is spanned by the member variables of the module, i.e. a state $s_o \in S_o$ is an n-tuple $s_o = (v^1, v^2, ..., v^n)$ with $n \in N$, where the v^i with $0 < i \leq n$ represent the individual member variables. The state of A_o can either be modified by A_b or A_t (indicated by the dotted arrows in Fig. 14.1).

Even though the current work does not explicitly deal with building hierarchical models, model checkers like NuSMV [2] have a notion of hierarchy and modules allowing to partition the above described FSMs in such a way that behavior, initiator, target, and object FSM of one SystemC module are grouped into one NuSMV module.

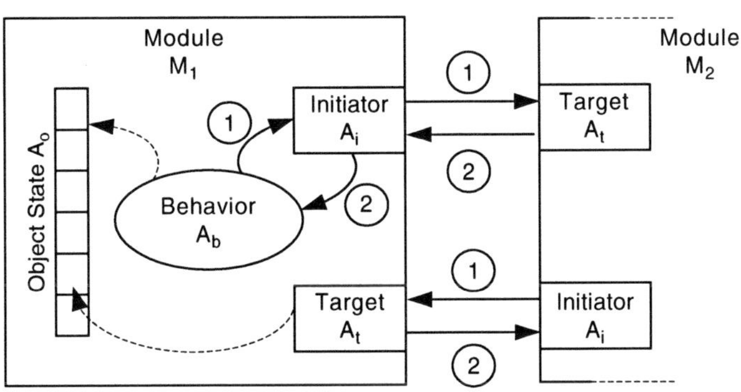

Figure 14.1. Model of a transaction level module composed of behavior, initiator, target, and object FSM. The solid arrows indicate transfer of control from one automaton to another, the numbers denote the order of the transfer. The dotted arrows show that the object state may only be changed by the behavior or target automaton.

Transfer of control between the FSMs. Let A_1 and A_2 be two FSMs. Transfer of control from A_1 to A_2 means that A_2 is in a state s_2^i, where only output ω_1^l of A_1 can trigger a transition, denoted by: $s_2^i \xrightarrow{?\omega_1^l} s_2^j$. After A_1 has generated output ω_1^l by a transition from s_1^l to s_1^m, formally written as $s_1^l \xrightarrow{!\omega_1^l} s_1^m$, it is in a state s_1^m with $s_1^m \xrightarrow{?\omega_2^k} s_1^n$, where only ω_2^k from A_2 can trigger a transition.

The solid arrows in Fig. 14.1 indicate a transfer of control from one automaton to another automaton. The numbers at the arrows indicate the order of the transfer. The behavior FSM, for example, transfers control to the initiator FSM, from where control is transfered back to the behavior FSM. It is important to note from Fig. 14.1 that A_t of module M_2 is waiting for an appropriate input from A_i of module M_1, while A_i itself is waiting for an appropriate input from A_b of M_1. Therefore, only the behavior FSM, or the process, may finally initiate a sequence of control transfers that constitutes a transaction.

The dotted arrows in Fig. 14.1 indicate that the object state may be changed from either the target FSM A_t or the behavior FSM A_b. Translated to SystemC, this means that the member variables of a module may either be changed by a process or an interface method of this module. This requires that the member variables of the module are not accessible from the outside world.

Example: Data flow source module. To discuss the basic concepts and illustrate our methodology, a simple source module of a data flow system is used in the following (see Fig. 14.2). The module issues blocking write transactions to a FIFO channel. The values written to the FIFO are incremented after each write; the usage of a three bit unsigned integer results in a wrap-around from seven to zero. The `sc_fifo<>` implementation from the SystemC reference implementation is used for the FIFO.

Behavior. Processes are modeled with the behavior FSM A_b. A state $s_b \in S_b$ is a two-tuple $s_b = (\nu, \lambda)$, where ν is a program counter identifying the statement to be executed and λ represents the local variables of the process. Common to every behavior automaton is that it has to remain in its initial state

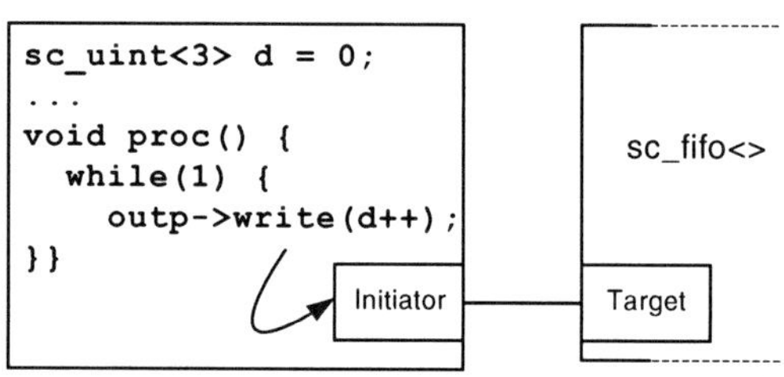

Figure 14.2. A data flow source module using blocking write.

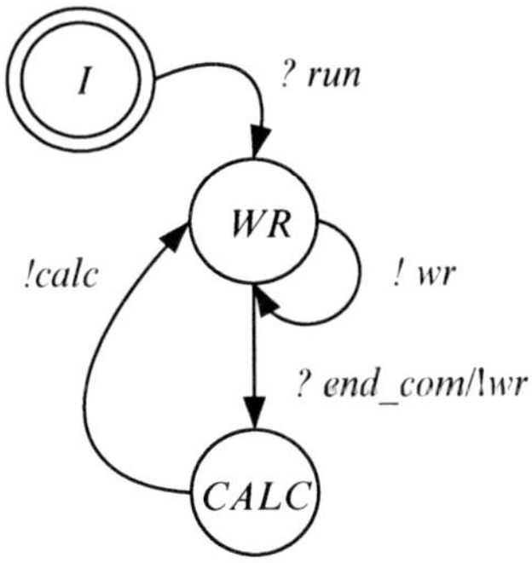

Figure 14.3. State diagram for the behavior FSM of the data flow module from Fig. 14.2.

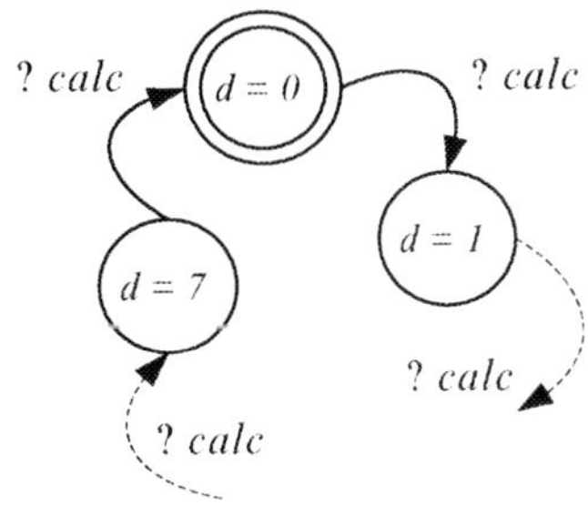

Figure 14.4. State diagram for the object FSM of the data flow model from Fig. 14.2.

until control is diverted to the process by the SystemC scheduler (for a detailed explanation of the scheduler see Section 5.0.0.0). The rest of the behavior FSM depends on the SystemC process to be modeled.

Figure 14.3 shows an example of state diagram of A_b for the module of Fig. 14.2. For the sake of simplicity, the program counter states are labeled with expressive names instead of numbers. Moreover, the process in the example does not have any local variables. The behavior FSM remains in its initial state I until input run is asserted by the simulation kernel. This models explicit selection of the running process by the scheduler. In state WR, output wr is generated and control is transfered to the initiator state machine; A_b will remain in state WR until the initiator FSM signals the end of communication with end_com. The initiator FSM A_i asserts output end_com whenever a transaction has been completed, thereby signaling to A_b that it can continue. After end_com has been received, state $CALC$ is entered from where an unconditional transition to WR is initiated. This transition generates an output $calc$.

Object state. The object state, represented by A_o is composed from all the member variables of the module. The internal state of a module may only be changed by a process or by calling a transaction implemented inside the module. Formally, this means that $\Sigma_o \subseteq \Omega_b \cup \Omega_t$, i.e. the input alphabet of A_o is a subset of union of the output alphabets of A_b and A_t.

In the example of Fig. 14.2, the only member variable is the three bit unsigned integer d. A transition of the object FSM A_o incrementing the value of d, see Fig. 14.4, is initiated at every occurrence of $calc$.

Initiator. Communication between the modules in a Transaction Level Model is modeled with an initiator FSM A_i and a target FSM A_t. We use $\langle T_i$ to denote the initiating state of a transaction T_i and $T_i\rangle$ for the terminating state

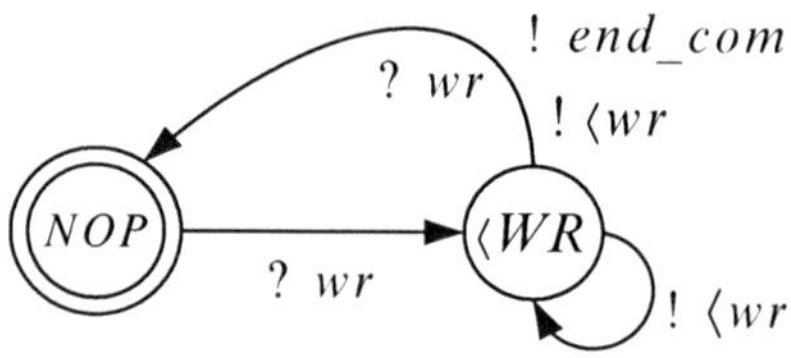

Figure 14.5. Initiator FSM, A_i, for the write transaction of the data flow example from Fig. 14.2.

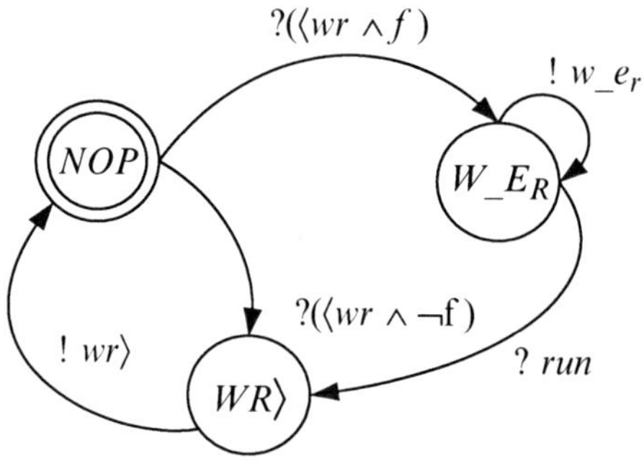

Figure 14.6. Target FSM, A_t, corresponding to the initiator FSM from Fig. 14.5. The target FSM implements the blocking write transaction requested by the initiator FSM.

of the transaction. The initiator FSM for the data flow example is shown in Fig. 14.5. From its initial state NOP it steps to the transaction initiator state $\langle WR$, if wr is asserted by A_b. In the transaction initiator state, the output $\langle wr$ is generated. The initiator FSM remains in $\langle WR$, until the end of the transaction is signaled from the target FSM A_t by $wr\rangle$. Finally, on leaving the transaction initiator state and returning to the initial state, end of communication is signaled to the calling process by output end_com. As each process may only request the initiation of one transaction at a time, it is sufficient to use one output end_com to indicate the end of all transactions that may be initiated by A_i.

Target. The target FSM A_t corresponding to the initiator FSM of Fig. 14.5 is shown in Fig. 14.6. It remains in the initial state NOP, until the execution of a transaction is requested by $\langle wr$. If the FIFO is full (f is $true$), the automaton proceeds to the blocking state W_E_r, modeling the blocking behavior of the write transaction. Note that the corresponding SystemC code to state W_E_r is a `wait()` statement, in which the process having called the write transaction is suspended until a *read event* e_r occurs, i.e. until at least one sample has been read from the FIFO. Any `wait()` statement can be modeled following the pattern described for state W_E_r. After having entered the blocking state W_E_r, output w_e_r is generated, signaling to the scheduler that the corresponding process should be suspended and only reactivated upon the occurrence of e_r.

The automaton remains in state W_E_r until run is asserted by the scheduler. The output run is generated by the scheduler after event e_r has been triggered and the process has been selected for execution (see Section 5). The end of transaction state, $WR\rangle$, can be reached in two ways. Either by leaving the blocking state W_E_r on the occurrence of run or directly from the initial state NOP, if the FIFO is not full. Upon transition from $WR\rangle$ to NOP, the end of transaction output, $wr\rangle$, is generated. Therefore, this transition initiates a transfer of control back to the initiator FSM.

5. Scheduling

To preserve the *cooperative multitasking* simulation semantics of SystemC [7] in the finite state model, a scheduler has to be included into our modeling effort. The scheduler in SystemC maintains a list of *runnable* processes and selects one of them nondeterministically to be *running*. This process is executed without interruption until either its end (*method processes*) or the occurrence of a `wait()` statement (*thread processes*). As every method process can be replaced by a functionally equivalent thread process, we will restrict our discussion to thread processes without loss of generality.

A process that has reached a `wait()` statement returns control to the scheduler, is removed from the list of runnable processes and is said to be *suspended*. Now, the scheduler selects another process from the list of runnable processes until the list is empty. The process of executing one runnable process after the other is called *evaluation phase*.

Once all processes are suspended, the scheduler calls the `update()` method of all *primitive channels* that have requested an update during the evaluation phase. A primitive channel is a channel that does not have its own processes or ports. Examples for primitive channels are hardware signals or FIFOs.

The next step of the scheduler is to check for *events* that have been *notified* during the evaluation phase and to add processes that are sensitive to one of the notified events to the list of runnable processes. Note that here we mean both kinds of sensitivity; *static sensitivity* (the event is in the sensitivity list) and *dynamic sensitivity* (the process waits for an event at a `wait()` statement embedded in the code). This phase is called *delta notification phase*.

The cycle of evaluation phase, update phase, and delta notification phase is called a *delta cycle*. If no processes are runnable at the end of a delta cycle, simulation time is advanced. Currently, we do not maintain a global simulation time, however, waiting for a specified amount of time and timed event notification is supported. Note that we deliberately do not support immediate notification, as it introduces nondeterminism into the design.

Modeling the Scheduler . Our model of the scheduler differs from the one presented in [9] by adding a notion of time and supporting dynamic sensitivity.

In [10] no clear separation between update phase and delta notification phase is made. The Petri net-based scheduler of [8] is closest to our model, however, they use a subscription scheme to associate processes and event notifications where we handle this association in the scheduler automaton.

A scheduler for N processes $P_0, P_1, ..., P_{N-1}$ is represented by an automaton A_s. It maintains a *process selector state* $\sigma \in \{none, P_0, ..., P_{N-1}\}$, for each process P_i a *process state* $\pi_i \in \{runnable, running, suspended\}$, and a global *phase selector state* $\phi \in \{evaluate, update, delta\_ntfy, timed\_ntfy\}$.

The scheduler is initialized to $s_s^0 = (\sigma = none, (\forall i \in \{0...N-1\} : \pi_i = runnable), \phi = evaluate)$, meaning that the scheduler starts in the evaluation phase with no process being run but all processes being runnable. Then, an arbitrary runnable process P_j is selected for execution, i.e. the process selector is set to $\sigma = P_j$ and the process state is changed to $\pi_j = running$. The process is executed until the occurrence of a wait statement after which it is suspended. The process state is changed to $\pi_j = suspended$ and the process selector is set to $\sigma = none$. This procedure is repeated until no process remains runnable, meaning until $\forall i \in \{0...N-1\} : \pi_i = suspend$ and $\sigma = none$.

Next, the scheduler switches to the update phase, $\phi = update$. All primitive channels receive an update signal in this phase. Each channel has the responsibility to decide whether it wants to execute its update method. Only after all updates have been carried out, the scheduler proceeds and changes the phase selector state to $\phi = delta\_ntfy$.

In the delta notification phase, events that have been notified during the evaluation phase are triggered. All processes that are sensitive (either statically or dynamically) to one of the triggered events, change their state from $suspended$ to $runnable$.

Now, the phase selector changes to $\phi = evaluate$ and the next delta cycle is started. If no processes become runnable at the end of a delta cycle, the phase selector is set to $\phi = timed\_ntfy$ and simulation time is advanced.

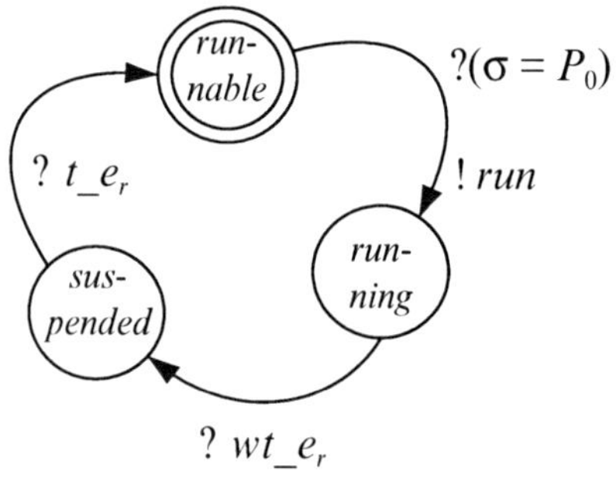

Figure 14.7. Process state for the example shown in Fig. 14.2.

In Fig. 14.7 the transitions of the process state for the example from Fig. 14.2 are shown. Once the process is selected for execution ($\sigma = P_0$), the process state changes from $runnable$ to $running$. Execution of the behavior automaton A_b is triggered by the assertion of run. The process remains in this state until the `wait()` statement in the blocking write transaction triggers a transition to $suspended$ by asserting w_e_r. Upon occurrence of t_e_r, the process state is changed to $runnable$ again. The input t_e_r corresponds to triggering event e_r. See Paragraph *Modeling Events and Time-Out* for details.

Modeling channel updates. The request-update mechanism for primitive channels requires an additional automaton A_u (see Fig. 14.8). The automaton starts in state rdy. On an update request req_up from within the channel, the automaton changes its state to $pend$. This happens during the evaluation phase of the scheduler. During the update phase of the scheduler, the automaton receives an input $update$ that initiates execution of the update functionality if the automaton is in state $pend$. The update functionality depends on the primitive channel being modeled. After having carried out the update, the automaton returns to its initial state, thereby signaling the end of the update.

Modeling events and time-out. Events are modeled by means of an automaton A_e (see Fig. 14.9). It has two states, $S_e = \{ntfy, cncl\}$, an input alphabet $\Sigma_e = \{\nu, \gamma, \delta\}$, an output alphabet $\Omega_e = \{nil, trig\}$, and an initial state $s_e^0 = cncl$. The occurrence of an event notification (a `notify` statement) is modeled by an input ν, a cancellation (a `cancel` statement) by γ. On a notification, the state is set to $ntfy$, on a cancellation to $cncl$. When the scheduler is in the delta notification phase and has added all processes sensitive to a notified event, to the list of runnable processes, the corresponding event automaton is reset by input δ. This assures that the same event notification does not trigger a process twice. The output of A_e is set to $trig$ if it is in state $ntfy$ and to nil if it is in state $cncl$.

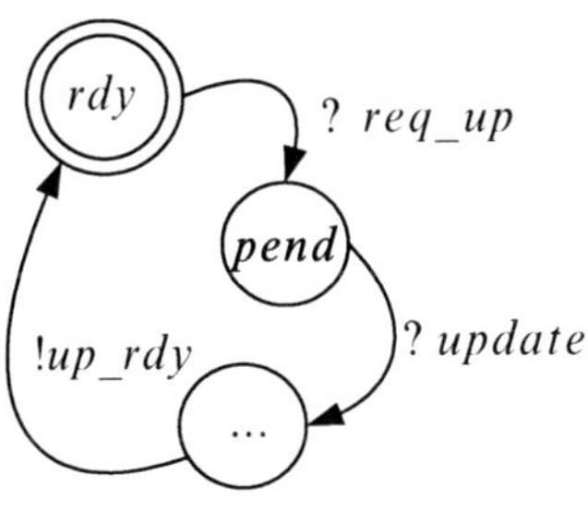

Figure 14.8. Generic state diagram for the update FSM.

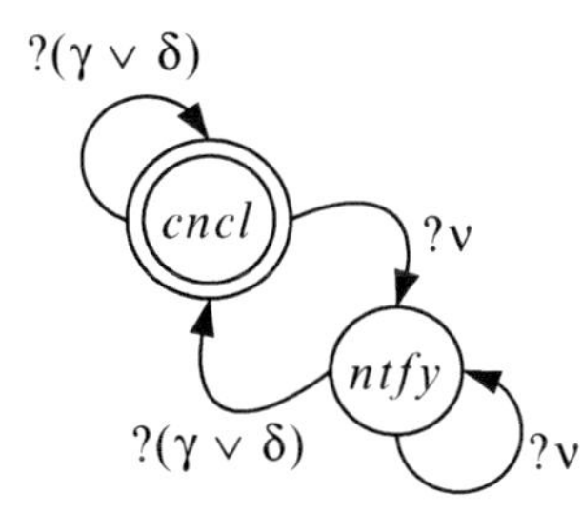

Figure 14.9. Event automaton.

Suspending a process for a certain amount of time, i.e. the `wait(<time>)` statement is supported with the help of a counter. At the occurrence of a timeout, the process state is changed from $running$ to $suspended$. Moreover, a counter is incremented each time the scheduler enters a timed notification phase, $\phi = timed\_ntfy$. If the counter has reached the value specified as an argument to the `wait(<time>)` statement, the process state is changed from $suspended$ to $runnable$ and the counter is reset to zero.

6. Experimental Results

In the following, we will demonstrate the applicability of our approach using several examples. All experiments were carried out using NuSMV [2] running on a 3.2 GHz Linux PC. For property specification, Computation-Tree Logic (CTL) has been used.

Writing properties. The advantage of the explicit modeling of initiator FSM, A_i, and target FSM, A_t, is the easy identification of transactions. A transaction T_i, for example, lasts as long as the output $\langle T_i$ of A_i is asserted. The end of a transaction T_i is marked with $T_i\rangle$. Stating in CTL that transactions T_1 and T_2 should never interfere, is as simple as $AG(\ \neg(\ \langle T_1 \wedge \langle T_2\ )\ )$. A CTL property that checks if a condition $cond$ in the target module holds after transaction T_1 has been executed can be formulated as $AG(\ T_1\rangle \rightarrow AX(cond)\ )$. As a last example, specifying that a blocking transaction T_1 initiated under a certain condition $cond$ should block until that condition is false, is stated as $AG((\ \langle T_1 \wedge cond\ ) \rightarrow A[\ \neg T_1\rangle\ U\ \neg cond\ ])$.

Traffic light system. A relatively simple example of a TLM is the traffic light system presented in [11]. A controller changes the status of two traffic lights by using transactions. We have modified the original model to have an additional red-yellow state. We have proven the usual safety and liveness properties:

P1 If one traffic light is green, the other one has to be red.

P2 Each traffic light has to be green infinitely often.

P3–P6 Verify the correct sequence of transaction calls, e.g. after a red light, red-yellow has to occur.

All properties could be verified within a fraction of a second.

Vending Machine. Another simple example is the vending machine presented in [12]. A user can insert a coin and select coffee or tea. This can be done in arbitrary order. After both transactions, inserting a coin and selecting a product, have been committed, the vending machine responds with delivering either coffee or tea to the user. The following properties were verified:

P1 After first inserting money and then selecting a product, the vending machine has to deliver the product.

P2 After first selecting a product and then inserting a coin, the vending machine has to deliver the product.

P3/P4 No product must be delivered without having inserted a coin/having selected a product.

Again, all properties could be verified within a fraction of a second.

Untimed data flow system. As a more complex example, we have implemented a simple data flow system. In its least complex version it consists of a source (SRC) connected to a sink (SNK) through a FIFO channel. In other versions, we have inserted a different number of feedthrough (FT) models between source and sink. The feedthrough models read data from the input FIFO and write it back to an output FIFO.

Source, sink, and feedthrough modules are implemented with `SC_THREAD` processes using blocking write and read access to the FIFO. The FIFO implementation we used is the `sc_fifo<>` primitive channel from the SystemC distribution. As we are mainly interested in communication, we have only modeled the number of samples stored into the FIFO and have abstracted from their values. The following properties were checked using different FIFO sizes:

P1/P2 The blocking write/read of the source always finally returns.

P3 A write to a full FIFO always blocks.

P4 A read from an empty FIFO always blocks.

The run-times needed to check those properties for different numbers of feedthrough modules and different FIFO sizes are shown in Table 14.1.

Timed data flow system. As an example of a timed, but unclocked system, we have modeled a different version of the data flow system using nonblocking read and write transactions. The source issues two consecutive nonblocking write accesses to the FIFO and then waits for T_0, while the sink always only issues one nonblocking read attempt before waiting for T_1. We assume that a non successful read or write attempt results in a loss of data. Therefore it is important to check whether every read or write attempt was successful. The properties that have been checked are summarized below:

P1/P2 Every write/read attempt was successful.

P3/P4 If the FIFO is full/empty, a write/read attempt fails.

Table 14.1. Run-times needed to check properties of the untimed data flow system. All times are measured in seconds. The size of the FIFOs is given in brackets, [].

		P1	P2	P3	P4
SRC-SNK	[64]	2 s	1 s	1 s	<1 s
SRC-SNK	[256]	366 s	62 s	11 s	11 s
SRC-FT-SNK	[16]	2 s	2 s	<1 s	<1 s
SRC-FT-SNK	[64]	1176 s	1452 s	4 s	4 s
SRC-FT-FT-SNK	[4]	95 s	115 s	2 s	2 s

Table 14.2. Run-times needed to check properties of the timed data flow system. All properties were verified for a FIFO size of 16. All times are measured in seconds.

	P1	P2	P3	P4
$T_0 = 40, T_1 = 20$	<1 s	<1 s	<1 s	<1 s
$T_0 = 40, T_1 = 21$	13 s	5 s	5 s	5 s

Table 14.2 shows the run-times measured for checking the different properties. Note, that property P1 for $T_0 = 40$ and $T_1 = 21$ fails because the source issues its write transactions with a higher frequency than the sink issues its read transactions.

Case study: simple bus system. The examples presented in the previous paragraphs are relatively simple and have mainly served as a pipe-cleaner during the development of the finite state formalism. They are very suitable for that purpose as they are easy to understand and can be used to study the basic transaction types — blocking and nonblocking — in a simple environment.

To demonstrate the applicability of the presented approach and to investigate limitations, a more elaborate example is needed. TLM, nowadays, is mostly used to model systems comprising several components communicating over a bus. A well-known, publicly available, reasonably complex TLM example is the *Simple Bus* system contained in the SystemC distribution. The Simple Bus is a cycle accurate model of a bus allowing multiple masters and multiple slaves to be connected to the bus. It supports single and burst requests. A request can either be a read or a write. The system contains models of a blocking master, a nonblocking master, an arbiter and two slaves, a fast memory, and a slow memory.

We have used the bus configuration shown in Fig. 14.10 as a case study. It consists of six modules. The bus itself is implemented inside the Simple Bus

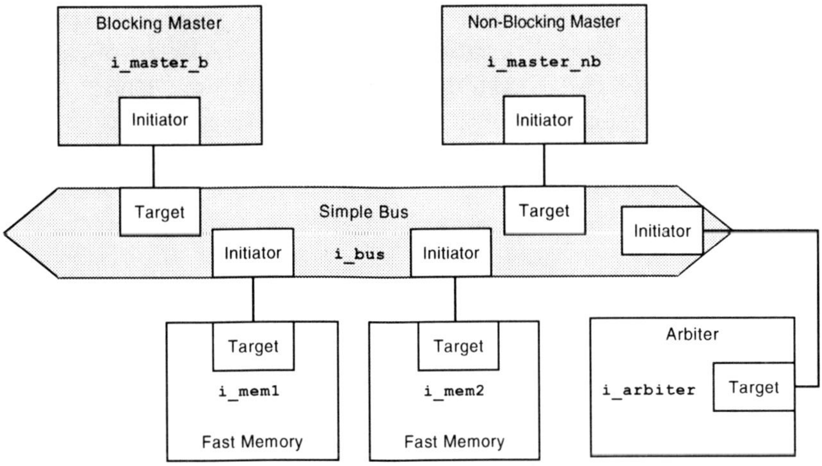

Figure 14.10. Block diagram of the simple bus system used as a case study. It has two masters, one blocking and one nonblocking, and two fast memories as slaves.

module. Two masters; the *Blocking Master* and *Non-Blocking Master*, can issue requests to the bus. Simultaneous or overlapping requests are handled by the *Arbiter*. The bus requests can have one of two *Fast Memory* slaves as targets. Three of the six modules have processes, namely the two masters and the bus itself. The two masters issue their requests at the positive clock edge, while the bus processes the requests at the negative clock edge. The model comprises three events, two of which are timed. The timed events are used to model time-outs between bus transactions issued by the masters. The untimed event signals the end of a transaction in the *Simple Bus* module. As communication over the bus is modeled at a cycle-accurate level, two additional timed events are needed, one to generate the positive clock edge, the other to generate the negative edge.

The SystemC code for the two masters, the bus, the arbiter, and the fast memory slave was converted to our finite state formalism. To reduce the translation effort, our finite state model does not support a lock mode for bus transaction which is present in the original SystemC model. The lock mode in the SystemC model prevents a burst request from being interrupted by another request. Moreover, to be able to apply model checking using NuSMV with reasonable run-times for the individual properties, some simplifications were made:

- We have abstracted from the data transfered over the bus.
- We have applied address abstraction, making use of the existence of an address map.

In the case of the simple bus, the first simplification is straight forward as the system behavior does not depend on the data values transfered over the bus. Abstracting from the data, therefore, does not introduce any nondeterminism. The introduction of an address abstraction does, however, introduce nondeterminism in some places. Instead of using integral address values, we only make a distinction whether an address value refers to the first or the second slave memory. Whenever it is not clear to which slave an address refers to, which, e.g. is the case after having applied an arithmetic operation on the address, one of the two slaves is chosen nondeterministically. This is similar to the address abstraction presented in [10].

We have checked more than 15 properties, some of which are listed below:

P1/P2 A read/write transaction from the nonblocking master always reaches the slave.

P3/P4 A read/write transaction from the blocking master always reaches the slave.

P5 The nonblocking master does not issue a request if the old request is still being processed.

P6/P7/P8 None of the three processes ever encounters a deadlock.

All properties could be verified within seconds up to several minutes, depending on the property.

The results presented so far are encouraging enough to motivate the application of the methodology to even more complex systems. This has been done by replacing one of the fast memories by a slow memory that takes several clock cycles to complete a read or write request. Unlike the fast memory, the slow memory contains its own process, increasing the total number of processes in the model from three to four. The resulting increase in run-time to more than one hour for some properties suggests that the simple bus system is an upper bound for the system complexity that can currently be handled by our methodology.

7. Conclusions

We have presented a finite state formalism for Transaction Level Models that allows to formulate properties based on transactions. We have shown the applicability of this approach with several simple examples and a more complex case study. It has been demonstrated that properties based on transactions can be formulated to prove the correctness of the system. Our methodology has three main advantages:

1. It is a native finite state model that does not need an additional translation step before being able to apply model checking. It can directly be used with existing model checkers like, e.g. NuSMV.

2. The emphasis is on the communication aspect, allowing convenient identification of transactions. Transactions can be used as atomic propositions within properties, thereby enabling a concise formulation of properties at a raised level of abstraction.

3. It can handle timed and untimed models at levels of abstraction above RTL.

Currently, the approach is mostly limited by the complexity of the behavior automata A_b used to model processes. The simple bus, for example, contains three processes and therefore three behavior automata. One is used to describe the bus itself, the other two are used to describe the blocking and the nonblocking master, respectively. As long as the two masters are simple and limited in their functionality, the system can be described using a finite state model. However, this will fail due to the state space explosion problem if the behavior processes in the masters contain complex algorithms and data dependent conditional statements. Also, programming techniques like dynamic memory allocation, recursion, or self modifying programs cannot be handled appropriately by the finite state approach.

Future work will therefore concentrate on abstraction methods for behavior automata in the masters. As our approach is focused on communication, abstraction, or simplification of the behavior automata in the communication resources is not desirable. Another direction for future work is to investigate possibilities to remove the explicit modeling of the SystemC kernel, as checking the correctness of the simulation kernel is not the objective of our model.

References

[1] Cai, L. and Gajski, D. (2003). Transaction level modeling: an overview. In: *CODES+ISSS '03: Proc. the 1st IEEE/ACM/IFIP International Conference on Hardware/Software Codesign and System Synthesis*, pp. 19–24, New York, ACM Press.

[2] Cimatti, A., Clarke, E., Giunchiglia, E., Giunchiglia, F., Pistore, M., Roveri, M., Sebastiani, R., and Tacchella, A. (2002). NuSMV 2: An opensource tool for symbolic model checking. In: *Proc. 14th International Conference on Computer-Aided Verification (CAV'2002)*.

[3] Dahan, A., Geist, D., Gluhkovsky, L., Pidan, D., Shapir, G., Wolfsthal, Y., Benalycherif, L., Kamdem, R., and Lahbib, Y. (2005). Combining system level modeling with assertion based verification. In: *Proc. the Sixth International Symposium on Quality Electronic Design (ISQED'05)*.

[4] Groe, Daniel and Drechsler, Rolf (2003). Formal verification of LTL formulas for SystemC designs. In: *IEEE International Symopsium on Circuits and Systems (ISCAS'03)*.

[5] Habibi, A. and Tahar, S. (2006). Design and verification of SystemC transaction-level models. *IEEE Transactions on Very Large Scale Integration (VLSI) Systems*, 14:57–68.

[6] Habibi, A., Tahar, S., Li, D., and Mohamed, O. A. (2006). Efficient assertion based verification using TLM. In: *Proc. Design, Automation and Test in Europe Conference and Exhibition (DATE'06)*.

[7] IEEE Computer Society (2006). IEEE Standard SystemC Language Reference Manual.
http://www.systemc.org.

[8] Karlsson, D., Eles, P., and Peng, Z. (2006). Formal verification of SystemC designs using a Petri-Net based representation. In: *Proc. Design, Automation and Test in Europe Conference and Exhibition (DATE'06)*.

[9] Kroening, D. and Sharygina, N. (2005). Formal verification of SystemC by automatic hardware/software partitioning. In: *Proc. MEMOCODE 2005*, pp. 101–110. IEEE.

[10] Moy, M., Maraninchi, F., and Maillet-Contoz, L. (2005). LusSy: a toolbox for the analysis of Systems-on-a-Chip at the transaction level. In: *Proc. the Fifth International Conference on Application of Concurrency to System Design (ACSD'05)*.

[11] Niemann, B. and Haubelt, C. (2006). Assertion-based verification of transaction level models. In: *Methoden und Beschreibungssprachen zur Modellierung und Verifikation von Schaltungen und Systemen*.

[12] Peranandam, Prakash M., Weiss, Roland J., Ruf, Juergen, and Kropf, Thomas (2004). Transaction level verification and coverage metrics by means of symbolic simulation. In: *Methoden und Beschreibungssprachen zur Modellierung und Verifikation von Schaltungen und Systemen*, pp. 260–269, Aachen, Shaker Verlag.

[13] Rose, A., Swan, S., Pierce, J., and Fernandez, J.-M. (2005). Transaction Level Modeling in SystemC. OSCI TLM Working Group.
http://www.systemc.org.

[14] Ruf, J., Hoffmann, D. W., Kropf, T., and Rosenstiel, W. (2001). Simulation-guided property checking based on multi-valued AR-automata. In: *Proc. Design Automation and Test in Europe (DATE)*, Munich.

Chapter 15

DIFFERENT KINDS OF SYSTEM DESCRIPTIONS AS SYNCHRONOUS PROGRAMS

Jens Brandt and Klaus Schneider
Reactive Systems Group
Department of Computer Science
University of Kaiserslautern
http://rsg.informatik.uni-kl.de

Abstract Many different system description and specification languages are used in modern design flows to emphasize different aspects like modular architecture, multithreaded behavior, abstract action-oriented behavior, and the desired temporal properties. However, the use of many specialized languages complicates the development of seamless and robust design flows. In this article, we show that synchronous languages are powerful enough to capture the mentioned aspects of system descriptions as simple syntactic sugar. In particular, we show how hardware structures, multithreaded and action-oriented programs as well as property specification languages can be incorporated in a synchronous programming language so that a single core language with a powerful compiler can handle all design descriptions in a consistent way.

Keywords Synchronous languages, specification languages, system design

1. Introduction

Many embedded systems are heterogeneous in the sense that they consist of parts that are currently developed by totally different design flows and languages. The most relevant distinction is thereby the partition into application-specific hardware and software. Different hardware architectures like DSPs or standard microprocessors moreover require different ways for the implementation of the software part, which often implies the use of specialized languages

S.A. Huss (ed.), Advances in Design and Specification Languages for Embedded Systems – Selected Contributions from FDL'06, 243–260.

and compilers. Various levels of abstraction make use of different modeling paradigms like action-oriented or multithreaded system behavior. Moreover, specialized languages are used to describe the desired functional and temporal properties for the simulation and verification phases. The set of languages used in modern system design therefore includes languages like system description languages as UML [29], SystemC [27], and SystemVerilog [2], hardware description languages like VHDL [40] and Verilog [26], property specification languages like PSL [1], and of course, traditional programming languages like C.

However, the plethora of languages currently used in many design flows is simply unmanageable. Many tools and licenses are required to update designs almost all the time; this makes the overall design process inefficient. Moreover, it is a big disadvantage that specialized languages force the designer to think already about a later realization (in software or hardware), which makes late design changes concerning the hardware–software partitioning or even weaker modifications of the architecture practically impossible. Specialized languages like Accellera's property specification language PSL [2] became moreover very complicated languages. However, the most difficult problem that results from the use of so many languages is that the underlying models of the languages do not always match: For example, hardware description languages usually rely on an event-based simulation, while software programs usually rely on a sequential uniprocessor execution, and property specifications consider formal models like transition systems.

For this reason, we propose in this article a unified approach that focuses on a single programming paradigm. This consolidation is the result of many bad experiences made with complex design flows. To this end, we propose the synchronous programming paradigm [3], since synchronous languages have already proved to be able to generate both application-specific hardware and software from the same synchronous program. Moreover, the underlying semantics match with the models used in formal verification like model checking.

However, synchronous languages currently do not offer much support for the temporal specifications. Usually, observers are written as synchronous programs that are then run in parallel with the system and report whether something bad has happened. Compilers for synchronous languages are able to traverse the state space and check whether such a situation can occur, so that checking safety properties is already comfortably embedded in synchronous programming environments. However, checking more complex specifications like liveness, different kinds of fairness, or even assume-guarantee reasoning is not yet available in commercial tools.

Based on the Esterel language, we developed our own experimental language called Quartz and implemented an entire tool framework called Averest for this language. The current version of Averest that is available under `www.averest.org` can be used to translate synchronous programs to hardware (netlists given in VHDL or Verilog) or software (programs written in C) [38, 37], and to verify given specifications written in temporal logics like LTL, CTL, and even in the full μ-calculus [36]. In this article, we present the new version of Averest that is currently in an experimental status, and therefore not yet publically available. During the development of this version, we also discussed additional special syntactic means to describe the modular architecture of a system, and also more powerful specifications to support development processes based on refinement techniques. However, having considered many languages and approaches, we came to the conclusion that most of these aspects can be implemented without much effort as simple syntactic sugar on top of our existing language.

This article therefore demonstrates the expressive power of the synchronous programming model in that we show how different aspects like the description of the modular architecture, complex properties as given by ω-regular properties similar to PSL and assume-guarantee reasoning can be incorporated in a synchronous language. In particular, we show that regular expressions and temporal logics can be easily translated to even more readable synchronous programs. Besides the better readability of the specification, the main advantage is the possibility to use existing tools for synchronous languages to simulate, verify, and debug the specified properties as well as the program. Moreover, all translations to hardware and software already offered by compilers for synchronous languages can still be used for hardware-software codesign.

The outline of the article is as follows: In the next section, we briefly consider the core of our synchronous language Quartz, which is the input language of our Averest system. In the following sections, we then show how structural descriptions, action-oriented descriptions, and property specifications are easily obtained in Quartz, so that these aspects can all be handled in a unique framework. For this reason, it is very simple to check their equivalence by means of verification or simulation.

2. The Averest System

Averest is a set of tools for specification, verification, and implementation of reactive systems. It includes a compiler for synchronous programs, a symbolic model checker, and a code generator for hardware and/or software synthesis. Further tools for simulation as well as for supervisor synthesis are currently under development. Averest can be used for modeling and verifying finite as well as infinite state systems at various levels of abstraction and with different kinds of descriptions. In particular, Averest is not only well suited for the design of integrated circuits, but also for developing communication protocols, concurrent programs, software in embedded systems, etc.

Systems descriptions for Averest are given in the imperative synchronous programming language Quartz [33–35, 37]. Quartz is an imperative synchronous language that was derived from the Esterel language [6, 5]. The common paradigm of synchronous languages is the perfect synchrony [17, 3], which means that most of the statements are executed as micro steps in zero time. Consumption of time is explicitly programmed by special statements which partition the micro steps into macro steps. In the programmer's view, all macro steps take the same amount of logical time. Thus, concurrent threads run in lockstep and automatically synchronize at the end of their macro steps. The introduction of micro- and macrosteps is not only a convenient programming model, it is also the key to generate *deterministic* single-threaded code from multithreaded synchronous programs. Thus, synchronous programs can be executed on ordinary microcontrollers without complex operating systems. As another advantage, the translation of synchronous programs to hardware circuits is straightforward [4, 32]. Moreover, the formal semantics of synchronous languages makes them particularly attractive for reasoning about program semantics and correctness. Therefore, synchronous languages are well suited for the design of safety-critical embedded systems that consist of application-specific hardware and software.

In this article, we mainly focus on the core of the Quartz language that is powerful enough to define many other statements as simple syntactic sugar. This core language consists of the following basic statements:

Definition 2.1. [Basic Statements of Quartz] The set of basic statements of Quartz is the smallest set that satisfies the following rules, provided that S, S_1, and S_2 are also basic statements of Quartz, ℓ is a location variable, x is an event variable, y is a state variable, σ is a Boolean expression, and α is a type:

- *nothing* (empty statement)
- $y = \tau$ and $\mathit{next}(y) = \tau$ (assignments)
- ℓ : *pause* (consumption of time)
- *if* (σ) S_1 *else* S_2 (conditional)
- $S_1; S_2$ (sequential composition)
- $S_1 \parallel S_2$ (synchronous concurrency)
- *do* S *while*(σ) (iteration)
- *[weak] suspend* S *when [immediate]*(σ) (suspension)
- *[weak] abort* S *when [immediate]*(σ) (abortion)
- $\{\alpha\ y;\ S\}$ (local variable y with type α)
- *choose* S_1 *else* S_2 (nondeterministic choice)
- *assume*(φ) (inline assumption)
- *assert*(φ) (inline specification)
- *name*$(\tau_1, \ldots, \tau_n)$ (module instance)

Many other statements can be defined as macro statements, e.g. the following *always* statement:

$$\textit{always } S :\equiv \textit{do } S; \textit{pause; while true}$$

In addition to many Esterel statements that can be defined in the above spirit as syntactic sugar, the Quartz language features moreover generic programs (compile time parameters), different forms of concurrency (synchronous, asynchronous, interleaved), explicit nondeterministic choice, fixed bitwidth integers with a complete set of arithmetic operations, arrays, infinite integers, and temporal logic specifications.

In Quartz, there are two kinds of (local and output) variables, namely *event* and *state variables*. State variables y are persistent, i.e., they store their current value until an assignment changes it. Executing a delayed assignment $\textit{next}(y) = \tau$ means to evaluate τ in the current macro step (environment) and to assign the obtained value to y in the following macro step. Immediate assignments update y in the current macro step and are therefore rather equations than assignments.

Event variables do not store their values, hence, an assignment $y = \tau$ to an event variable just gives y the value τ for this moment of time (unless there is another assignment in the next instant with the same value). If no assignment takes place, the value is not stored, instead, a default value is taken. As most events are of Boolean type, we use the statements *emit* x and *emit* $\textit{next}(x)$ as macros for $y = \text{true}$ and $\textit{next}(y) = \text{true}$, respectively.

There is only one basic statement that defines a control flow location, namely the *pause* statement.[1] For this reason, we optionally[2] endow *pause* statements with unique Boolean valued *location variables* ℓ that are true iff the control is currently at location ℓ : *pause*.

The semantics of the other statements is essentially the same as in Esterel. Due to lack of space, we do not describe their semantics in detail, and refer instead to [37] and, in particular, to the Esterel primer [5], which is an excellent introduction to synchronous programming.

Moreover, Quartz supports *generic programming* in that sequences, and parallel statements can be described via compile-time parameters. For example, it is therefore possible to implement a system with n similar threads or with n-bit wide variables, or with arrays of length n. Moreover, statements can be given by primitive recursion so that n *if-then-else* statements can be nested in each other.

In addition to the above statements, there are many more statements that are, however, simply reduced to the above core language. For example, using oracle inputs with nondeterministic choice, different kinds of concurrency like asynchronous and interleaved concurrency can be easily reduced to synchronous concurrency.

```
module BehaveDETECT110(event i,&o)
implements SpecDETECT110(i,o) {
   event prv_i, prv_prv_i;
   loop {
      if(i) emit next(prv_i);
      if(prv_i) emit next(prv_prv_i);
      if(!i&prv_i&prv_prv_i) emit o;
      pause;
   }
}
```

Figure 15.1. Behavioral description of a 110-detector.

Quartz programs are simply a list of modules, where only the first (main) module is considered by the compiler. The other modules are either instantiated in the first module or they are simply not used. A Quartz module consists of a header that contains the name of the module, the declaration of the inputs and outputs,[3] as well as the body statement. A very simple example module is shown in Fig. 15.1 that describes a module that checks whether the boolean-valued input stream i contains 110 as a subsequence. In case this subsequence has been read, the output o is made true.

3. Structural Descriptions

Structural descriptions emphasize the hierarchy that is obtained by the composition of modules to new modules at the next level of the hierarchy. Structural descriptions are particularly popular in classic hardware design. In Quartz, module instances can be used as ordinary statements, so that instances can run in sequence or in parallel, and can interact with each other.

It is therefore straightforward to describe modular hardware structures by Quartz modules. To this end, no special syntax is necessary, all that is required is a restriction to certain statements: *Basic structural modules*, which form the leaves of the hierarchy, are constructed by a behavioral description, i.e. their modules' bodies contain a Quartz statement. Figure 15.2 shows some simple definitions of basic hardware gates.

Composed structural modules simply consist of a local declaration of the internal variables (wires in the case of hardware circuits), and a parallel execution of the instantiated hardware modules. It might seem to be wastefull to run a single thread for each hardware gate, but this may not be the case in the generated code: The compiler is able to merge all the loops that are obtained by expansion of the modules, so that a single thread can simulate the gate netlist in software (if wanted). A similar argument holds for hardware code generation, so that

```
module AND(event a,b,&out) {
   always
      if(a&b) emit out;
}

module OR(event a,b,&out) {
   always
      if(a|b) emit out;
}

module NEG(event a,&out) {
   always
      if(!a) emit out;
}

module DFF(event a,&out) {
   always
      if(a) emit next(out);
}
```

Figure 15.2. Behaviors of basic hardware gates.

the number of location variables is reduced to only one pause statement (all the location variables are easily detected to have the same transition relations, so that the compiler can unify them all).

An example is shown in Fig. 15.3, where we implemented a hardware circuit for detecting a 110 subsequence in the input stream. As can be seen, structural descriptions are naturally obtained in a synchronous language like Quartz. Moreover, no special treatment is required for the compiler: it is obviously still possible to generate hardware and software from these modules. In case of hardware design, this offers the benefit of a highly efficient simulation at the synchronous level by compiling the obtained C programs individually for the particular design. For hardware synthesis, the compiler essentially generates the gates that have been used in the description. Hence, there is no loss of the hardware structure and neither a loss of efficiency due to the compilation.

4. Action Languages

Action languages are frequently used for modeling software or hardware systems at an abstract level. Examples of action languages are Unity [11], DisCo [22], TLA [24], sublanguages used in UML[4] as well as languages used in model checkers like Murphi [14] or ALF [9].

```
module DETECT110_Structure(event i,&o) implements
DETECT110_Spec(i,o){
   event w1, w2, w3, w4, w5, w6, w7;
   {
      NEG(i,w1);
   || AND(i,w7,w2);
   || DFF(w2,w3);
   || NEG(w7,w4);
   || AND(i,w4,w5);
   || DFF(w5,w6);
   || OR(w3,w6,w7);
   }
}
```

Figure 15.3. Structural implementation of a 110-detector.

The general model of computation of these languages is thereby that a program consists of a set of actions that are executed whenever an associated condition holds. In this sense, this computation model is related to the "guarded commands" that were already considered by Dijkstra [12].

The translation of Quartz programs consists of computing also a set of guarded commands [38]: The compiler computes for each assignment $x = \tau$ the guard condition, i.e. the condition that holds iff the assignment is executed. Then, for each variable x, the hardware code generation will generate a case construct that checks the different guards and assigns the corresponding right hand sides.

It is therefore straightforward to directly integrate action languages in the Quartz language. The compiler is thereby even simplified, since the conditional actions can be directly used for the intermediate data structures of the compiler. Moreover, it is very simple to implement a given set of conditional actions as a Quartz module: given actions (γ_1, α_1), ..., (γ_n, α_n), we simply use the following module:

```
module ModuleName(...){
  always{
    if(γ1) α1;
    ⋮
    if(γn) αn;
  }
}
```

The above scheme is very simple, but nevertheless very powerful. In particular, the combination of other Quartz statements like parallel execution, abortion, suspension, etc. allows us to capture also complex module transformations [23] like merging of action modules and many others.

5. Property Specification

In this section, we consider the main ingredients of property specifications available in Quartz. Quartz offers additional constructs to specify complex temporal properties which we have to omit due to lack of space. In particular, temporal logics like CTL, LTL, special fragments of CTL*, as well as the full μ-calculus can be used for this purpose. In particular, past temporal operators are often very convenient as can be seen even with the simple example given in Fig. 15.4: `s1` means that at all points of time (`G`) on all computation paths (`A`) of the system, `o` holds iff currently `i` is false, and `i` was true at the preceeding two points of time. `s2` is another temporal specification stating the same property with only future operators. Finally, `s3` considers only one implication, and can therefore make use of the CTL logic and its more efficient model checking algorithms.

Besides the possibility to simply list temporal logic specifications as shown in the specification module given in Fig. 15.4, Quartz offers also inline specifications that refer to the corresponding control flow location of the program. These inline specifications are similar to those developed for VHDL in [31]. Moreover, it is possible to make use of regular and ω-regular expressions similar to Accellera's industry standard property specification language PSL. In contrast to PSL, we make use of Quartz statements to replace regular expressions by more readable program statements.

In the remainder of this section, we first present the possibilities of inline specifications and assumptions to support readable specifications as well as assume-guarantee reasoning. Then, we show how regular expressions can be translated to Quartz statements, and finally we briefly discuss similar translations for temporal logics.

```
spec SpecDETECT110(event i,&o) {
   s1 : A G (o <-> !i & PSX (i & PSX i));
   s2 : A G (i & X i & X X !i <-> X X o);
   s3 : A G (i -> A X (i -> A X (!i -> o)));
}
```

Figure 15.4. Temporal specification of a 110-detector.

Inline Assertions and Assumptions

Verifying a system aims at ensuring that the system has exactly the specified behavior. Traditionally, two kinds of specifications can be distinguished: *White box specifications* may refer to internal states or local variables of a module, while *black box specifications* only describe the external behavior of a module without referring to internals. In general, black box specifications should be preferred to support a hierarchical reasoning that is independent of a particular implementation. However, a specification may not hold all the time, but only after reaching some point of the computation which is conveniently defined with inline (white box) specifications.

In traditional programming languages, which lack a built-in verification support, programmers are used to employ assertions for this task: They specify a condition that should hold at a particular control flow location by means of *assertion statements*. Whenever the control flow hits such a statement during the execution of the program, the given condition is checked and the program is aborted if the condition is violated.

Quartz supports both black box and white box specifications by means of *assume* and *assert* statements. Like any statement, they can be used at an arbitrary point of a Quartz module and both statements take a list of labelled conditions, where the conditions can be given in the temporal logic LTL. However, their semantics is completely different: *Assume statements list conditions that the programmer assumes to hold at the corresponding control flow location, whereas assert statements list conditions that have to be checked at that location.* The verification tool trusts the programmer and uses the assumptions to check the other conditions. The task of assume statements is to give the verification tool additional information that can not be derived from the program alone. For example, knowledge about the environment and possible input values or complex mathematical relations can be added in this way. In many cases, it is possible to check the given assumptions if the final context is given by other modules, but some assumptions are directly given by the environment and can therefore not be checked unless a model of the environment is provided.

Of course, assume guarantee reasoning [19, 28, 41] is directly supported by the assume and assert statements. Moreover, statements in a module can be annotated with classical pre/postconditions and loop invariants [15, 20, 13]. The task of the compiler is the extraction of the given assumptions and proof obligations and their hand-over to a model checker. Additionally, it creates specifications for common program errors: For example, for bitvectors, it is checked whether an overflow occurs due to arithmetic expressions, and for arrays, it is checked that the index expressions remain in the declared bounds.

Assume/assert statements implement a white box specification. Specification of the black box behavior is achieved by defining a *specification module* (an example is shown in Fig. 15.4). The body of a specification module does not consist of a statement; instead, a list of temporal specification is given that have to be fulfilled by any concrete module that *implements the specification module*. The *implements* clause in the header of a concrete module tells the compiler to set up also proof obligations to check the temporal properties of the referred specification module.

Abstract modules are again only syntactic sugar: In principle, they consist of a sequence of assert statements that list the temporal specifications. For code generation, assume and assert statements are implemented in an observer that will set an error flag in case the property is violated during run-time.

In this way, Quartz allows the programmer the step-wise refinement of a given system. Starting with a pure temporal specification, proceeding with an intermediate action-oriented description, and finally concluding with a behavioral Quartz statement, a programmer can refine a given model of a system so that each refinement step can be proved to be correct. Moreover, as Averest is capable to handle unbounded integers, one can first start with unbounded integers, then one could check how large the values may grow to finally determine sufficient bitwidths. Hence, also the datatypes can be refined during this process.

Regular Expressions and Finite Automata

In the previous section, we have shown where specifications can be placed in a module, but we left open, which logic is used to write down the formal specifications. Quartz allows one to use temporal logic specifications, in particular, the linear temporal logic LTL and the branching time temporal logic CTL. Moreover, certain fragments of the more powerful logic CTL* are considered, and even the full μ-calculus is supported, which is currently one of the most expressive specification logics [36].

We discussed also the use of Accellera's property specification language PSL for future versions of Averest, but came to the conclusion that synchronous languages can be used to make such specification much more readable: While PSL also provides LTL and CTL, it additionally considers regular expressions and finite as well as infinite paths in order to support both simulation and verification, respectively.

Regular expressions can also be added to Quartz as syntactic sugar: It is well known that regular expressions, right- and left-linear grammars, and finite automata in different variants are equivalent to each other [21, 7, 8, 18, 10, 30]. Regular expressions are a very simple, yet powerful formalism to describe languages that can be accepted by finite state automata. Note that regular expressions describe languages of finite words over a fixed alphabet Σ. For the specification of reactive systems, however, also infinite computations have to be taken into account, which can be done by ω-regular expressions [39].

In the following, we demonstrate that, using our language Quartz, we are able to write powerful specifications that are as readable as programs. We believe that a programming approach to specification is more appreciated by programmers and engineers. To this end, we will consider a finite fixed set of variables $\mathcal{V}$. The alphabet $\Sigma_{\mathcal{V}}$ of $\mathcal{V}$ is then simply the powerset of $\mathcal{V}$, hence, $\Sigma_{\mathcal{V}} :\equiv 2^{\mathcal{V}}$. Then, regular expressions are defined as follows:

Definition 5.1 (Regular Expressions)**.** The set of general regular expressions $\mathsf{RegExp}(\Sigma)$ over the alphabet Σ is defined as the smallest set that satisfies the following properties (where $\alpha, \beta \in \mathsf{RegExp}(\Sigma)$):

- $\emptyset \in \mathsf{RegExp}(\Sigma)$ and $1 \in \mathsf{RegExp}(\Sigma)$
- $\vartheta \in \mathsf{RegExp}(\Sigma)$ with $\vartheta \subseteq \Sigma$ (letters)
- $\alpha + \beta \in \mathsf{RegExp}(\Sigma)$ (union)
- $\alpha \& \beta \in \mathsf{RegExp}(\Sigma)$ (intersection)
- $\alpha\beta \in \mathsf{RegExp}(\Sigma)$ (concatenation)
- $\alpha^* \in \mathsf{RegExp}(\Sigma)$ (finite iteration)

The above definition of regular expression contains already some syntactic sugar. The semantics of a regular expression r is a set of finite words $\mathsf{Lang}(r) \subseteq \Sigma_{\mathcal{V}}^*$ that is recursively defined as follows (ε is the empty word):

- $\mathsf{Lang}(\emptyset) := \{\}$ and $\mathsf{Lang}(1) = \{\varepsilon\}$
- $\mathsf{Lang}(\vartheta) := \{\vartheta\}$ for every letter $\vartheta \subseteq \Sigma$
- $\mathsf{Lang}(\alpha + \beta) := \mathsf{Lang}(\alpha) \cup \mathsf{Lang}(\beta)$
- $\mathsf{Lang}(\alpha \& \beta) = \mathsf{Lang}(\alpha) \cap \mathsf{Lang}(\beta)$
- $\mathsf{Lang}(\alpha\beta) := \{ab \mid a \in \mathsf{Lang}(\alpha), b \in \mathsf{Lang}(\beta)\}$
- $\mathsf{Lang}(\alpha^*) := \bigcup_{i=0}^{\infty} \mathsf{Lang}(\alpha)^i$

It is well known how to translate a given regular expression r to a finite state automaton $\mathcal{A}_r$ that accepts the language $\mathsf{Lang}(r)$ [21, 8, 18, 10, 30]. It is even possible to compute a symbolic representation of such an automaton in time $O(|r|)$ of length $O(|r|)$ [36]. As Quartz programs are also a way to symbolically describe automata, we can directly translate regular expressions to equivalent Quartz statements, which is done by the following function $\mathbb{Q}$:

- $\mathbb{Q}(\emptyset) :\equiv \texttt{assert(false)}$
- $\mathbb{Q}(1) :\equiv \texttt{assert(true)}$
- $\mathbb{Q}(\vartheta) :\equiv \texttt{assert}\ \varphi_\vartheta; \texttt{pause};$
  with $\varphi_\vartheta :\equiv \left(\bigwedge_{a\in\vartheta} a\right) \wedge \left(\bigwedge_{a\notin\vartheta} \neg a\right)$ for $\vartheta \subseteq \mathcal{V}$
- $\mathbb{Q}(\alpha \& \beta) :\equiv \mathbb{Q}(\alpha) \parallel \mathbb{Q}(\beta)$

```
loop
   choose {
      assert(a&!b)
      pause;
   } else {
      assert(a&!b);
      pause;
      assert(!a&b);
      pause;
   }
```

Figure 15.5. Quartz program for $(\{a\} + \{a\}\{b\})^\omega$.

- $\mathbb{Q}(\alpha + \beta) :\equiv$ `choose` $\mathbb{Q}(\alpha)$ `else` $\mathbb{Q}(\beta)$
- $\mathbb{Q}(\alpha\beta) :\equiv \mathbb{Q}(\alpha);\ \mathbb{Q}(\beta)$
- $\mathbb{Q}(\alpha^*) :\equiv$ `finloop` $\mathbb{Q}(\alpha)$
- $\mathbb{Q}(\alpha^\omega) :\equiv$ `loop` $\mathbb{Q}(\alpha)$

`finloop` is thereby a loop that only finitely often iterates its body statement, but the number of iterations is nondeterministic. Using nondeterministic Quartz statements, it is possible to implement such a loop together with a temporal assertion statement:

```
finloop S :≡  {event d;
               choose emit d else nothing;
               while(d){
                 S;
                 assert Fℓ;
                 choose emit d else nothing;
               }}
               ℓ : pause;
```

As an example, consider the regular expression $(\{a\}+\{a\}\{b\})^\omega$ that describes all infinite words over the alphabet $\{\{\}, \{a\}, \{b, \}, \{a, b\}\}$ that does not contain two succeeding occurrences of b. It is translated to the Quartz program given in Fig. 15.5.

Observers for Temporal Logic

Although Quartz allows one to use full temporal logic formulas[5] as specifications, it is sometimes more convenient to write observers with some reachability or fairness constraints. This is the classic approach that is used in many synchronous programming environments. Moreover, it is straightforward to compute for temporal past properties an equivalent deterministic (!) finite state automaton that can be used as an observer [36]. It is even straightforward to implement temporal past operators by means of equivalent Quartz modules that can then be simply called to "execute" the specification.

```
module PastAlways(event phi, &failure) {
   while(phi) {
      pause;
   }
   emit failure;
}

module PastUntil(event phi,psi, &failure) {
   bool q;
   q = false;
   loop {
      next(q) = psi | phi & q;
      pause;
   }
}
```

It is well known how to translate LTL formulas φ to equivalent ω-automata $\mathcal{A}_\varphi$, and even translations to symbolic or alternating automata exist that work in linear time (see [36] for further references). We can also use these algorithms to directly generate nondeterministic Quartz statements with assert statements to capture the fairness requirements that are generated by these translations.

6. Summary and Conclusions

In this article, we have shown that different kinds of system descriptions can be obtained on the basis of a unique programming paradigm. We use the synchronous programming paradigm to describe modular concurrent systems like hardware gate netlists, action-oriented modules, as well as temporal assertions (even extended by ω-regular expressions) in form of simple syntactic sugar. Hence, synchronous programs may not only serve as realization independent descriptions for either hardware or software; they can also be used as alternatives to complex property specification languages like PSL. In addition to increased readability, the approach offers also simulation and execution of these specifications on different architectures.

Notes

1. To be precise, immediate forms of suspend also have this ability.
2. In case the programmer does not provide a location variable, the compiler will automatically generate one.
3. Outputs are distinguished from inputs by prefixing their name with a & symbol.
4. see www.omg.org
5. Full temporal logic includes also past time operators, which makes the logic not more powerful, but exponentially more succinct and in general more readable [16, 25, 36].

References

[1] Accellera (2004a). Property specification language reference manual, version 1.1. http://www.eda.org.

[2] Accellera (2004b). SystemVerilog 3.1a language reference manual. Technical report. Accellera's Extensions to Verilog.

[3] Benveniste, A., Caspi, P., Edwards, S., Halbwachs, N., Le Guernic, P., and de Simone, R. (2003). The synchronous languages twelve years later. *Proc. IEEE*, 91(1):64–83.

[4] Berry, G. (1991). A hardware implementation of pure Esterel. In: *Workshop on Formal Methods in VLSI Design*, Miami, Florida.

[5] Berry, G. (2000). The Esterel v5_91 language primer.

[6] Berry, G. and Gonthier, G. (1992). The Esterel synchronous programming language: design, semantics, implementation. *Science of Computer Programming*, 19(2):87–152.

[7] Berry, G. and Sethi, R. (1986). From regular expressions to deterministic automata. *Theoretical Computer Science*, 48(1):117–126.

[8] Brüggemann-Klein, A. (1992). Regular expressions into finite automata. In: Simon, I., editor, *Latin American Symposium on Theoretical Informatics (LATIN)*, vol. 583 of *LNCS*, pp. 87–98, São Paulo, Brazil, Springer.

[9] Bultan, T. (2000). Action language: A specification language for model checking reactive systems. In: *International Conference on Software Engineering (ICSE)*, pp. 335–344, University of Limerick, Ireland, IEEE Computer Society.

[10] Champarnaud, J.-M. (2001). Implicit structures to implement NFA's from regular expressions. In: Yu, S. and Paun, A., editors, *Conference on Implementation and Application of Automata (CIAA)*, vol. 2088 of *LNCS*, pp. 80–93, London, Ontario, Canada, Springer.

[11] Chandy, K.M. and Misra, J. (1989). *Parallel Program Design*. Addison-Wesley, Austin, Texas.

[12] Dijkstra, E.W. (1975). Guarded commands, nondeterminacy, and formal derivation of programs. *Communications of the ACM*, 18(8):453–457.

[13] Dijkstra, E.W. (1976). *A Discipline of Programming*. Prentice-Hall.

[14] Dill, D.L. (1996). The Murphi verification system. In: Alur, R. and Henzinger, T.A., editors, *Conference on Computer Aided Verification (CAV)*, vol. 1102 of *LNCS*, pp. 390–393, New Brunswick, NJ, Springer.

[15] Floyd, R.W. (1967). Assigning meaning to programs. In: *Symposia in Applied Mathematics: Mathematical Aspects of Computer Science*, vol. 19, pp. 19–31.

[16] Gabbay, D.M., Pnueli, A., Shelah, S., and Stavi, J. (1980). On the temporal analysis of fairness. In *Symposium on Principles of Programming Languages (POPL)*, pages 163–173, New York, ACM Press.

[17] Halbwachs, N. (1993). *Synchronous programming of reactive systems.* Kluwer.

[18] Halbwachs, N., Lagnier, F., and Raymond, P. (1993). Synchronous observers and the verification of reactive systems. In: *Conference on Algebraic Methodology and Software Technology (AMAST)*, Workshops in Computing, pp. 83–96, Twente, Springer.

[19] Henzinger, T.A., Qadeer, S., and Rajamani, S.K. (1998). You assume, we guarantee: methodology and case studies. In: Hu, A.J. and Vardi, M.Y., editors, *Conference on Computer Aided Verification (CAV)*, vol. 1427 of *LNCS*, pp. 440–451, Vancouver, BC, Canada, Springer.

[20] Hoare, C.A.R. (1969). An axiomatic basis for computer programming. *Communications of the ACM*, 12:576–580.

[21] Hopcroft, J.E. and Ullman, J.D. (1979). *Introduction to Automata Theory, Languages and Computation.* Addison-Wesley.

[22] Järvinen, H. and Kurki-Suonio, R. (1990). The DisCo language and temporal logic of actions. Technical Report 11, Tampere University of Technology, Software Systems Laboratory.

[23] Kurki-Suonio, R. (2005). *A Practical Theory of Reactive Systems – Incremental Modeling of Dynamic Behaviors*. Springer.

[24] Lamport, L. (1991). The temporal logic of actions. Technical Report 79, Digital Equipment Cooperation.

[25] Markey, N. (2003). Temporal logic with past is exponentially more succinct. *Bulletin of the European Association for Theoretical Computer Science*, 79:122–128.

[26] Moorby, P. (1992). History of Verilog. *IEEE Design and Test of Computers*, pp. 62–63.

[27] Open SystemC Initiative (2003). SystemC 2.0.1 language reference manual. http://www.systemc.org.

[28] Pasareanu, C.S., Dwyer, M.B., and Huth, M. (1999). Assume-guarantee model checking of software: a comparative case study. In: Dams, D., Gerth, R., Leue, S., and Massink, M., editors, *Model Checking Software (SPIN Workshop)*, vol. 1680 of *LNCS*, pp. 168–183, Toulouse, France, Springer.

[29] Rational Software Corporation (1997). UML Notation Guide - version 1.1.

[30] Raymond, P. and Roux, Y. (2002). Describing non-deterministic reactive systems by means of regular expressions. *Electronic Notes in Theoretical Computer Science (ENTCS)*, 65(5). Workshop on Synchronous Languages, Applications, and Programming (SLAP).

[31] Reetz, R., Schneider, K., and Kropf, T. (1998). Formal specification in VHDL for formal hardware verification. In: *Design, Automation and Test in Europe (DATE)*. IEEE Computer Society.

[32] Rocheteau, F. and Halbwachs, N. (1991). Pollux, a Lustre-based hardware design environment. In: Quinton, P. and Robert, Y., editors, *Conference on Algorithms and Parallel VLSI Architectures II*, Chateau de Bonas.

[33] Schneider, K. (2000). A verified hardware synthesis for Esterel. In: Rammig, F.J., editor, *Workshop on Distributed and Parallel Embedded Systems (DIPES)*, pp. 205–214, Schlo<C3><9F> Ehringerfeld, Germany, Kluwer.

[34] Schneider, K. (2001). Embedding imperative synchronous languages in interactive theorem provers. In: *Conference on Application of Concurrency to System Design (ACSD)*, pp. 143–156, Newcastle upon Tyne, UK, IEEE Computer Society.

[35] Schneider, K. (2002). Proving the equivalence of microstep and macrostep semantics. In: Carre<C3><B1>o, V., Mu<C3><B1>oz, C., and Tahar, S., editors, *Higher Order Logic Theorem Proving and its Applications (TPHOL)*, vol. 2410 of *LNCS*, pp. 314–331, Hampton, VA, Springer.

[36] Schneider, K. (2003). *Verification of Reactive Systems – Formal Methods and Algorithms*. Texts in Theoretical Computer Science (EATCS Series). Springer.

[37] Schneider, K. (2006). The synchronous programming language Quartz. Internal Report (to appear), Department of Computer Science, University of Kaiserslautern.

[38] Schneider, K., Brandt, J., and Schuele, T. (2006). A verified compiler for synchronous programs with local declarations. *Electronic Notes in Theoretical Computer Science (ENTCS)*, 153(4):71–97.

[39] Thomas, W. (1990). *Automata on Infinite Objects*, vol. B, Automata on infinite objects, pp. 133–191. Elsevier.

[40] VHDL93 (1993). *IEEE Standard VHDL Language Reference Manual.* New York. ANSI/IEEE Std 1076-1993.

[41] Zulkernine, M. and Seviora, R.E. (2002). Assume-guarantee algorithms for automatic detection of software failures. In: Butler, M.J., Petre, L., and Sere, K., editors, *Integrated Formal Methods (IFM)*, vol. 2335 of *LNCS*, pp. 89–108, Turku, Finland, Springer.

IV

UML-BASED SYSTEM SPECIFICATION AND DESIGN

Introduction

This fourth part of the book is a selection of the most interesting results in the thematic area 'UML-Based System Specification & Design' This FDL'06 thematic area addresses specification and design methodologies such as the Model Driven Architecture (MDA) approach, that use UML (Unified Modeling Language) to map abstract models of complex embedded systems to highly programmable hardware platforms and heterogeneous systems on chip.

The first three papers in this part of the book have been presented in the session 'Design Flows for Systems-on-Chip'. The first paper 'A Model-driven co-design flow for Embedded Systems,' by Elvinia Riccobene, Patrizia Scandurra, Alberto Rosti and Sara Bocchio, reports on results of the realization of a model driven design path from UML to C/C++/SystemC. The second paper describes 'A Method for Mobile Terminal Platform Architecture Development' and is by Klaus Kronlöf, Samu Kontinen, Ian Oliver and Timo Eriksson. This paper presents a method for developing service oriented platform architectures starting from end user requirements. The aim is to develop executable models suitable for stakeholders validation. The third paper, 'UML2 Profile for Modeling Controlled Data Parallel Applications' by Ouassila Labbani, Jean-Luc Dekeyser, Pierre Boulet and Èric Rutten proposes a UML solution for modeling control automata that can be applied in parallel systems.

The last two papers selected for this part of the book have been presented in the session 'Models for Design Space Exploration'. 'MCF: A Metamodeling based Visual Component Composition Framework' by Deepak Mathaikutty and Sandeep Shukla describes a framework for architecture exploration on various levels of abstraction. The last paper in this part is not least. It received the FDL'06 best paper award and is by Henk Corporaal, Marcel Verhoef, Oana Florescu, Jeroen Voeten. 'Reusing Real-Time Systems Design Experience Through Modelling Patterns' addresses exploration of real-time properties of architectures. Patterns enable easy composition of alternative architectures.

Piet van der Putten
Department of Electrical Engineering
Technische Universiteit Eindhoven
Eindhoven, The Netherlands
p.h.a.v.d.putten@tue.nl

Chapter 16

A MODEL-DRIVEN CO-DESIGN FLOW FOR EMBEDDED SYSTEMS

Sara Bocchio[1], Elvinia Riccobene[2], Alberto Rosti[1], and Patrizia Scandurra[2]
[1] *STMicroelectronics Lab R&I, Agrate Brianza, Italy*
{sara.bocchio; alberto.rosti}@st.com
[2] *University of Milan, Dipartimento di Tecnologie dell'Informazione, Crema, Italy*
{riccobene; scandurra}@dti.unimi.it

Abstract The UML (Unified Modeling Language), with the enhancements in UML2, is receiving interest by an increasing number of industrial and academic groups from the embedded software and hardware areas, who look at it and at its extension mechanisms as a practical and standard means to define family of languages targeted to specific application domains and levels of abstraction, while providing unification. In the Embedded Systems and SoC (System-on-Chip) area, we defined a model-driven design methodology based on UML 2.0, UML profiles and C/C++/SystemC. In this chapter, we extend this design flow in order to support the *platform-based design* principles. We also present the architecture of a prototype tool, which provides a graphical representation in UML (from a high-level functional model down to RTL) of HW and SW components, C/C++/SystemC code generation from UML models, and a reverse engineering process from C/C++/SystemC code to UML.

Keywords Embedded systems, SoC design, UML, SystemC, model-driven engineering (MDE), model-driven architecture (MDA), refinement

1. Introduction

System-on-Chip (SoC) design may involve the mixing on a single integrated circuit of one or more microprocessor cores (e.g. ARM, MIPS.), bus interface, analog components, and numerous digital processing functions. Designers are increasingly reusing portions of previous designs to reduce the time to market which generally results in greater revenue for the product. In the past few years, major functions have been implemented as virtual components [5]. To minimize

S.A. Huss (ed.), Advances in Design and Specification Languages for Embedded Systems – Selected Contributions from FDL'06, 265–284.

effort and risk in a new design, some organizations are internally standardizing on a set of virtual components and any related software to develop their own SoC platforms. Platform-based design allows an organization to develop a complete SoC that is central to its product line. Once the SoC platform is fully operational, derivative designs in which only a few virtual components are added or dropped are accomplished rapidly.

In [33, 12], the authors identify two layers for platform definitions: the micro-architecture platform and the application programming interface (API) platform. A micro-architecture platform is defined as a specific family of micro-architectures, oriented toward a particular class of problems, which can be modified (extended or reduced) by the system developer. API platform usually consists of a software layer that wraps the essential parts of the architecture platform and defines the services that the platform offers. It includes, among other things, RTOS and device drivers, that can be preferably modeled as a separated layer. In this chapter, we present a systematic approach through UML 2.0 for assembling micro-architecture platform and for defining the connection among the micro-architecture, the API, and the application tasks.

In the embedded design community, UML 2.0 and its extension mechanism are receiving significant interest as standard approach to define *family of languages* targeted to specific application domains and levels of abstraction. This is confirmed by current standardization activities controlled by the OMG such as: the Schedulability, Performance, and Timing Analysis (SPT) profile [27]; the recent UML for SoC Forum (USoC) [32] in Japan founded by Fujitsu, IBM, and CATS to define a set of UML extensions to be used for SoC design; the SysML proposal [28] which extends UML toward the Systems Engineering domain, and the recent MARTE (Modeling and Analysis of Real-Time Embedded systems) profile initiative [26]. Moreover, several reported experiences and contributions to the theme *UML for Embedded Systems* exist in literature (see [11], [13], [7]).

In accordance with the OMG *Model-driven Architecture* (MDA) [14] – a framework for Model-driven Engineering (MDE) [3, 4] – in [22] and [1] we defined a model-driven SoC design methodology which involves the UML 2.0, a UML 2.0 profile for the C and the SystemC language [18, 29], and some other UML profiles primarily related to the SW implementation. Here we extend this design flow in order to support the *platform-based design* principles in [33] [12], an important paradigm for the future embedded system development based on the reuse of generic hardware platforms.

This chapter is organized as follows. In Section 2, we describe the design flow, while in Sections 3 and 4, we present the two UML profiles for the HW and SW descriptions, respectively. In Section 5, we describe the environment architecture and its components features that we developed to assist the designer across the refinement steps in the UML modeling activity starting from a high-level functional model of the system down to RTL. In Section 6, we discuss some case studies. In Section 7, we quote some related work. Finally, in Section 8, we sketch some future directions.

2. The Model-driven Design Flow

Figure 16.1 summarizes the most significative phases of our new design flow. Essentially, the UML – or the Systems Modeling Language (SysML) [28] or the MARTE proposal [26] – in a platform-independent manner is used as schematic entry to provide a first specification of the system in terms of an *executable model* suitable to perform high-level functional validation and eventually performance analysis.

At this point, an available generic hardware platform for the physical architecture is selected and then the mapping – dictated by the sense and experience of expert engineers – of the software components on the given hardware platform is done.

This mapping phase is a meet-in-the-middle process. It is carried out through an iterative activity. The platform model is configured appropriately according to a given HW/SW partitioning, and then executed to check whether the QoS requirements (delays, power consumption, hardware resources, real-time, and embedding constraints) imposed by the system requirements are satisfied. The result is a particular platform instance with software embedded (the *Embedded System platform model*) at a specific level of abstraction (functional untimed/-

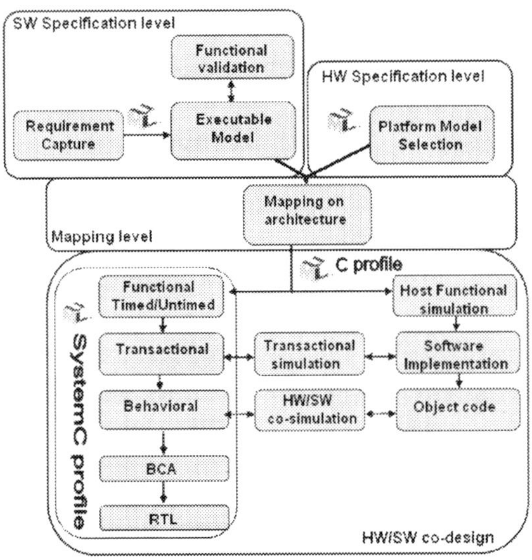

Figure 16.1. A model-driven design flow for ES.

timed, transactional, behavioral, bus-cycle accurate, RTL) depending on the amount of details of the provided platform model and on the description level of the SW components.

At UML level, the mapping of the application model on the given platform model is intended as *model weaving*, i.e. an operation which establishes semantic links between models at specific *joint points*. As input, this task requires also a reference model of the mapping (the *Mapping model* or *Weaving model*) to try, which is specified in terms of UML component and deployment diagrams to denote and annotate the partitioning of the original system in HW and SW components. This mapping model establish the relationships (joint points) of the platform resources and services with the application-level functional components. The components assigned to the HW partition are mapped directly onto a HW resource (i.e. a HW node in the UML deployment diagram) of the micro-architecture platform. The components in the SW partition are implemented by the SW designer as SW tasks that use the services provided by the API platform; this is reflected at UML level by the interface usage/realization dependencies between the application components and the components of the API platform.

After mapping, two different design flows start for the software and the hardware parts respectively. The hardware platform is modeled at different levels of abstraction (*functional untimed/timed*, *transactional*, *behavioral*, *bus cycle accurate*) on top of the RTL level by constructs from the UML profile for SystemC; so it can also be refined down to the RTL, modeling the details of the implementation platform.

The application can be modeled in software at three levels of abstraction: functional, transactional, or instructional level. This does not necessarily imply a software refinement path, but just a mapping on more detailed platforms to analyze better the performances of the overall application. At functional level, the C application is divided in threads running on a *host* machine – **Host Functional simulation**: in this first phase the SW designer decides the thread partition according to the inner application parallelism. At transactional level, the application is modeled using the same UML profile for SystemC. In this case, it works as a library encapsulated in a SystemC module; processes are associated to the software functional description to sustain its concurrent activity within the system, whereas communication is implemented by transactions that model the interactions with the hardware architecture – **Transactional simulation**. This level allows testing the correctness of the application on a high level description of the hardware, focusing on the communication and performances. Finally, the application can be also represented at instruction level by integrating an instruction set simulator (ISS) of the target processing unit within the SystemC environment, to execute the compiled application code together with the model of the hardware – **cycle-accurate HW-SW co-simulation**.

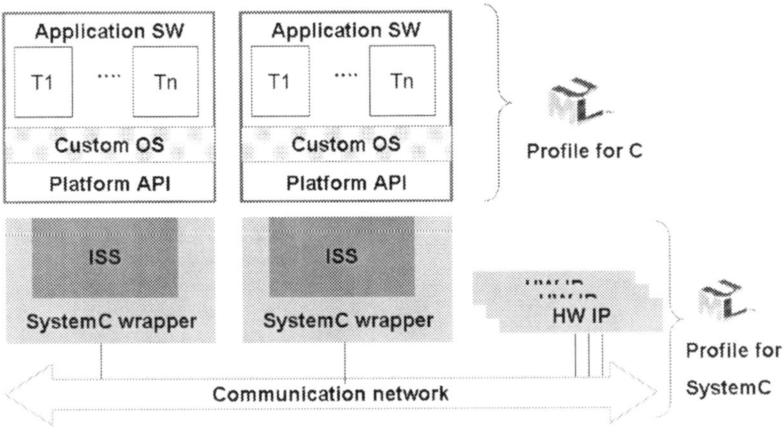

Figure 16.2. Platform model.

In this last case, the application software can be viewed as organized in different layers on top of the micro-architecture platform, as shown in Fig. 16.2. The lower layer provides the driver and the architecture controller: this layer is the application programming interface (API) platform and it is typically provided by the platform designer. The upper layer are the OS and the SW application developed by the SW designer.

3. The UML Profile for the HW

The UML 2.0 profile for SystemC [21] is a consistent set of modeling constructs designed to lift both structural and behavioral features of the SystemC language (including events and time features) to UML level. Based on the UML 2.0 specification [31] and on the SystemC 2.0 specification [29], the profile is defined at two distinct levels – the SystemC core layer (or layer 0) and the SystemC layer of predefined channels, ports, and interfaces (or layer 1) – which reflect the layered-architecture of SystemC. The complete UML profile definition for SystemC is described in [21, 24].

The core layer – *the basic SystemC profile* – is the foundation upon which specific libraries of model elements or also other modeling constructs can be defined. It is logically structured to reflect the core layer (or layer 0) of SystemC. A **Structure and Communication** part defines stereotypes for the primitive building blocks of SystemC like modules, interfaces, ports and channels. Fig. 16.3 shows an example of a SystemC module having a multiport, an array port, and a simple port, together with the port type and interface definitions of the simple port.

UML class diagrams are used to define modules, interfaces, and channels. The internal hierarchical structure of composite modules, especially the one of

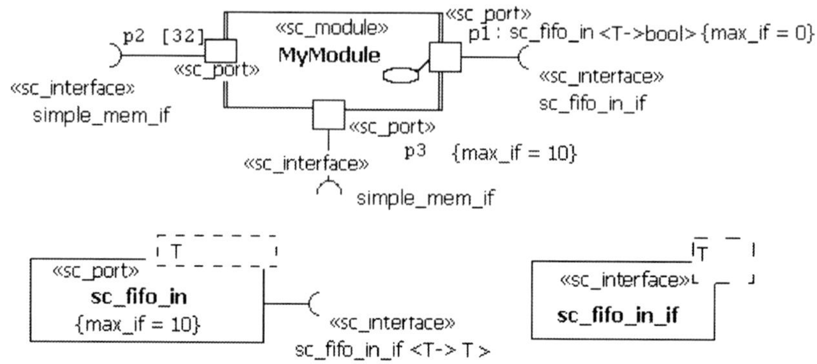

Figure 16.3. Examples of ports.

the topmost level module (which represents the structure of the overall system), is captured by UML composite structure diagrams; then, from these diagrams several UML object diagrams can be created to describe different configuration scenarios. This separation allows the specification (also partial) of different HW platforms as instances of the same parametric model (i.e. the composite structure diagram).

A **Behavior and Synchronization** part defines special state and action stereotypes which lead to a variation of the UML state machine diagram, the *SC Process State Machines*. This formalism has been appositely included in the profile definition to model the control flow and the reactive behavior of SystemC processes (methods and threads) within modules. A finite number of *abstract behavior patterns* of state machines [20] have been identified and can be used for modeling. Fig. 16.4 depicts one of these behavior patterns for a thread that: (i) is not initialized, (ii) has both a static (the event list $e_{1s}, \ldots, e_{Ns}$) and a dynamic sensitivity (the `wait` state), and (iii) runs continuously (by the infinite `while` loop).

Figure 16.5 shows a module class `count_stim` containing a thread process `stimgen`, two input ports `dout` and `clock`, and two output ports `load` and `din`. In particular, `clock` is a behavior port since it provides events that trigger the `stimgen` thread process within the module. The `stimgen` thread state machine is shown in Fig. 16.6. It is a realist example of the behavior pattern presented in Fig. 16.4.

A **Data types** part defines a UML class library to represent the set of SystemC data types. In addition, the predefined channels, ports and interfaces of the layer 1 of SystemC are considered. These concepts are provided either as a UML class library, modeled with the basic stereotypes of the SystemC core layer, or as a group of stand alone stereotypes – *the extended SystemC profile* – which specializes the basic profile. Currently, we have been working to include also

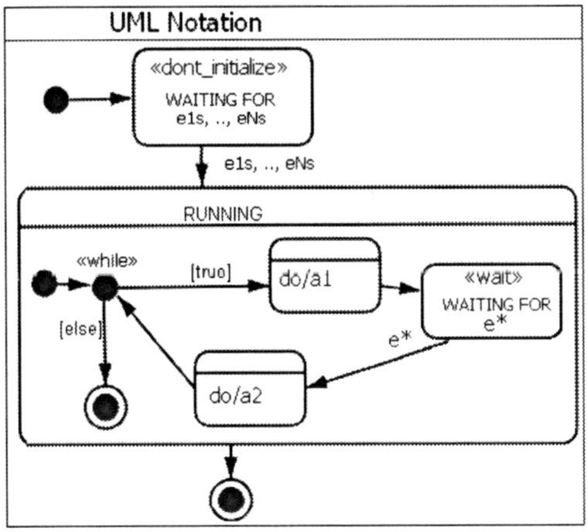

Figure 16.4. A thread process pattern.

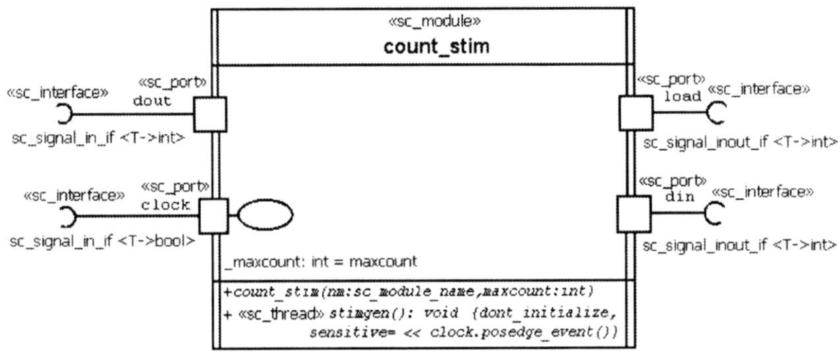

Figure 16.5. Example of a module, its attributes, ports, functions, and processes.

the **OSCI TLM 1.0 library** of channels and interfaces [18], in order to provide a UML class library for modeling at the TLM level of abstraction.

Using SystemC to link the system level SoC design flow to the consolidated VLSI design flow is a well-known issue. What is innovative is the idea to rely on low-cost customized (by a standard profiling technique) UML visual modeling tools in the early stages of the design process as front ends of lower level HW/SW codesign frameworks (these last would be used therefore for the final exploration and synthesis only).

Our goal is raising the level of abstraction to develop more complex systems by manipulating models only along the refinement steps from a high functional level down to RTL. To give an idea on how to perform refinement at UML level by the use of the SystemC UML profile, below we present an example of model refinement.

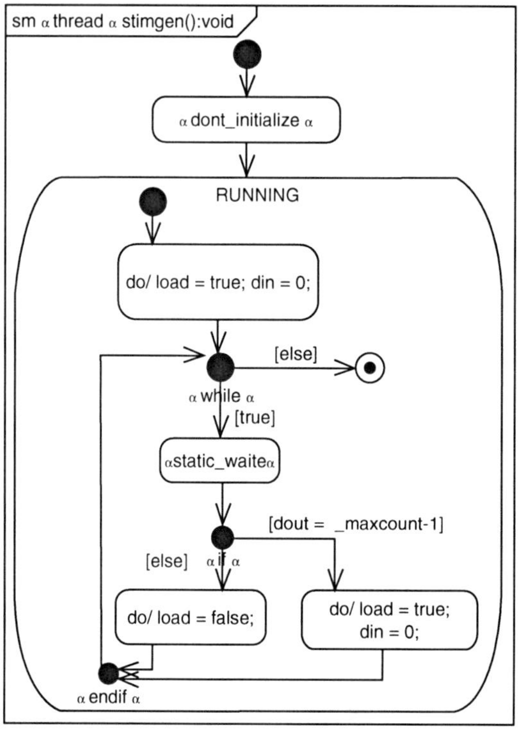

Figure 16.6. A thread process state machine.

A refinement example. We show here how to apply a refinement strategy to refine the communication aspects of a simple abstract (functional timed) producer/consumer system taken from [29]. In general, in the refinement process we not only refine the model's internal structures, its timing, or the datatypes being used; we also need to think about how components communicate with their environment. *Communication refinement* refers to mapping an abstract communication protocol into an actual implementation related to a given target architecture.

Before taking a closer look at the specific example, we clarify the process of communication refinement in a more general way. To this purpose, we assume to have two high level modules M1 and M2 communicating over a channel C via some abstract protocol. As suggested in [29], one possible approach to refining this basic communication scenario toward an implementation consists of the following steps:

1. Select an appropriate communication scheme to implement
2. Replace the abstract communication channel C with a refined one $C_{Refined}$ which realizes the selected communication protocol

3. Enable the communication of the modules M1 and M2 over $C_{Refined}$ by either: (a) wrapping $C_{Refined}$ in a way that the resulting channel $C_{Wrapped}$ provides the interfaces required by M1 and M2 (***wrapping***), or (b) refining M1 and M2 into $M1_{Refined}$ and $M2_{Refined}$, respectively, in order their required interfaces match the ones provided by $C_{Refined}$ (***adapter-merging***)

The two cases (not the only ones) are similar. Both include an intermediate step to build two further modules (i.e. two SystemC hierarchical channels), say *adapters*, to map one interface to another: one, say A1, between M1 and $C_{Refined}$, and one, say A2, between $C_{Refined}$ and M2. In the case (a), the resulting channel $C_{Wrapped}$ encloses $C_{Refined}$ and the two adapters; while, in the case (b) these adapters are merged to the calling modules M1 and M2 resulting in the refined modules $M1_{Refined}$ and $M2_{Refined}$. Deciding whether to use wrapping or merging depends on the methodology and chosen target architecture.

We can now come back to our specific example. The design of a working producer/consumer module that writes and reads characters to/from a FIFO channel is shown by the UML composite structure diagram in Fig. 16.7. The top composite module is defined to contain one instance of the consumer module, one instance of the producer module, and one FIFO channel instance. The FIFO channel permits to store characters by means of blocking read and write interfaces, such that characters are always reliably delivered. Two processes, the producer and the consumer (see the thread processes main within the producer and the consumer modules in Fig. 16.8), respectively feed and read the FIFO. The producer module writes data through its out port into the FIFO by a sc_fifo_out_if interface, the consumer module reads data from the FIFO through

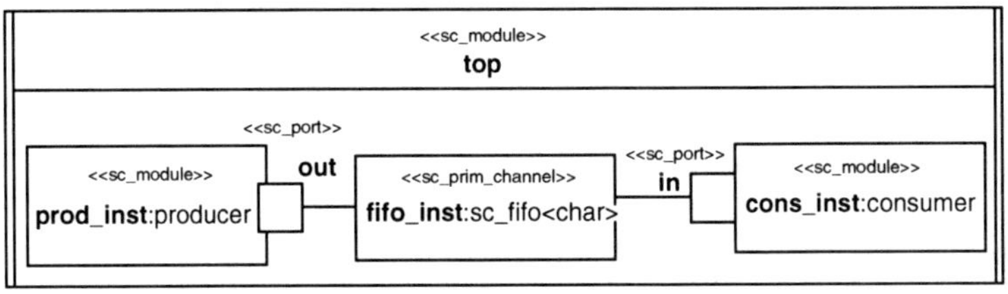

Figure 16.7. A producer/consumer design.

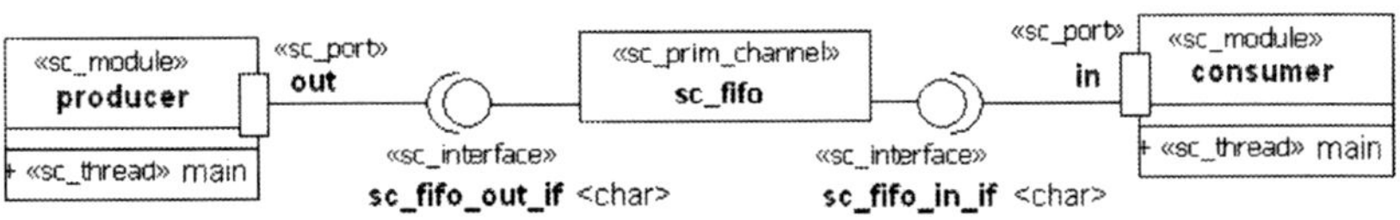

Figure 16.8. Producer/consumer modules communicating via a primitive FIFO.

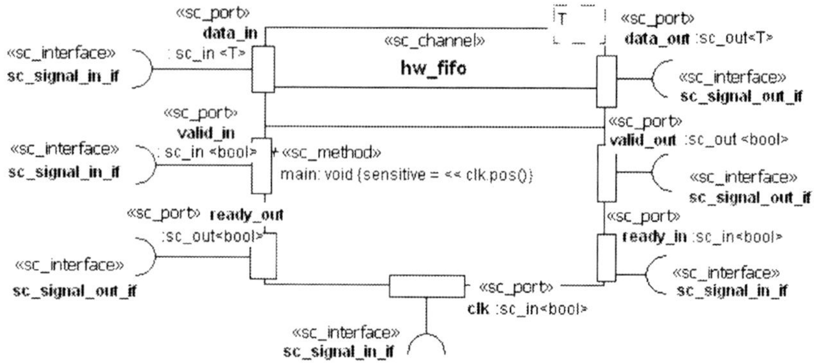

Figure 16.9. Clocked RTL HW FIFO.

its in port by the sc_fifo_in_if interface. These two interfaces are implemented by the FIFO channel (see the sc_fifo channel in Fig. 16.8). Because of the blocking nature of the sc_fifo read/write operations, all data are reliably delivered despite the varying rates of production and consumption.

Now, let us assume to replace the (abstract) FIFO instance above with a (refined) model of a clocked RTL hardware FIFO named hw_fifo<T> for an hardware implementation. The new hardware FIFO uses a signal-level ready/-valid handshake protocol for both the FIFO input and output (see Fig. 16.9). It should be noted that we cannot use hw_fifo directly in place of sc_fifo, since the former does not provide any interfaces at all, but has ports that connect to signals, i.e. has ports that use the sc_signal_in_if and sc_signal_out_if interfaces.

Following the wrapper-based approach (3.b) of the refinement procedure described earlier, we can define a hierarchical channel hw_fifo_wrapper<T> ($C_{Wrapped}$) which implements the sc_fifo_out_if and sc_fifo_in_if interfaces and contains an instance of hw_fifo<T> ($C_{Refined}$). In addition, it contains sc_signal instances to interface with hw_fifo<T> and a clock port (since hw_fifo<T> has also a clock port) to feed in the clock signal to the hw_fifo<T> instance (see Fig. 16.10). Finally, we need to add a hardware clock instance in the top-level design to drive the additional clock port that is now on the hw_fifo_wrapper<T> instance (see Fig. 16.11). The hw_fifo_wrapper<T> implements the required signal-level ready/valid handshake protocol whenever a read or write operation occurs; this protocol will properly suspend read or write transactions if hw_fifo<T> is not ready to complete the operation. Details on the SystemC code can be found in [29].

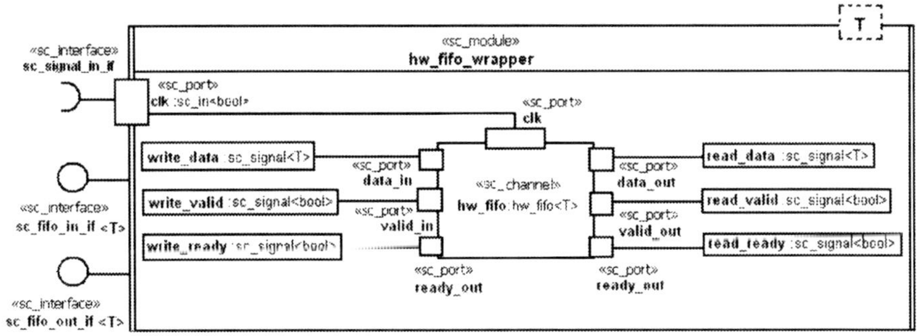

Figure 16.10. The hw_fifo_wrapper hierarchical channel.

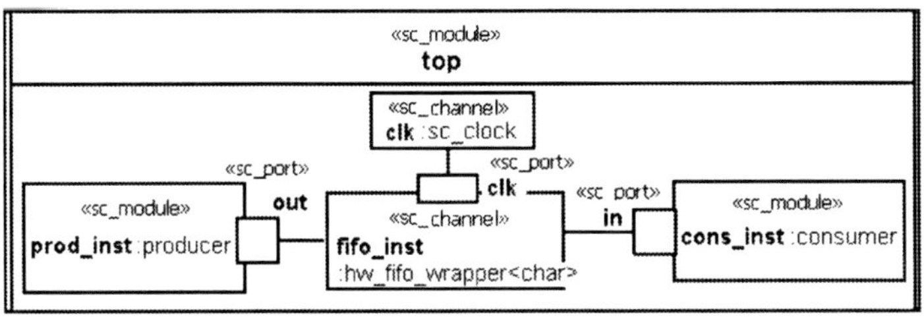

Figure 16.11. A clocked producer/consumer design.

4. The UML Profile for the SW

We model the software by using a UML profile for a multithread C/C++ programming language. We describe the structure of the software application by class diagrams, eventually exposing interfaces. The software architecture can be better described by composite structure diagrams. A class is considered *active* if it contains a thread that generates activity within the model. In the profile we introduced the stereotype thread to denote active threads as specialized class operations. For the description of the threads behavior, the UML profile for C/C++ supports a state-chart formalism which is similar to the one of the SystemC UML profile, but with some simplifications to take away sensitivity mechanisms like wait states and other SystemC-specific concepts.

The SW application model can be executed at different levels of abstraction. The target platforms are basically three:
(i) the *host*, i.e. a Linux platform – it is used to provide a fast functional validation of the SW application only;
(ii) SystemC – a transactional level model of the application runs within the SystemC environment and interacts with the hardware parts of the architecture platform model;

(iii) a *multiprocessor* platform, made of a few instances of ISSs – a compiled version of the application runs on a set of ISSs which can interact with hardware parts of the architecture platform model. Depending on the target platform, we produce therefore a functional implementation running on Linux, a transactional SystemC model, or an instruction level model for ISSs. We choose among the possible implementations by drawing a stereotyped mapping dependency between active classes and the components representing the target platform (see Fig. 16.12). The reactive behavior of a thread within an active class of the application model is modeled by a state machine diagram and admits alternative implementations. On Linux the thread is mapped onto a Linux thread, on SystemC it is mapped on a SystemC thread, and on the ISS it is mapped on a thread handled by the OS layer.

As in the UML profile for SystemC, the communication among functional parts is modeled through ports and interfaces which can be easily translated into function calls to the API platform. The communication can be also modeled through shared variables in the scope of a class or of a common ancestor class; in this case the API platform provides the mechanism to implement a mutual exclusion access policy.

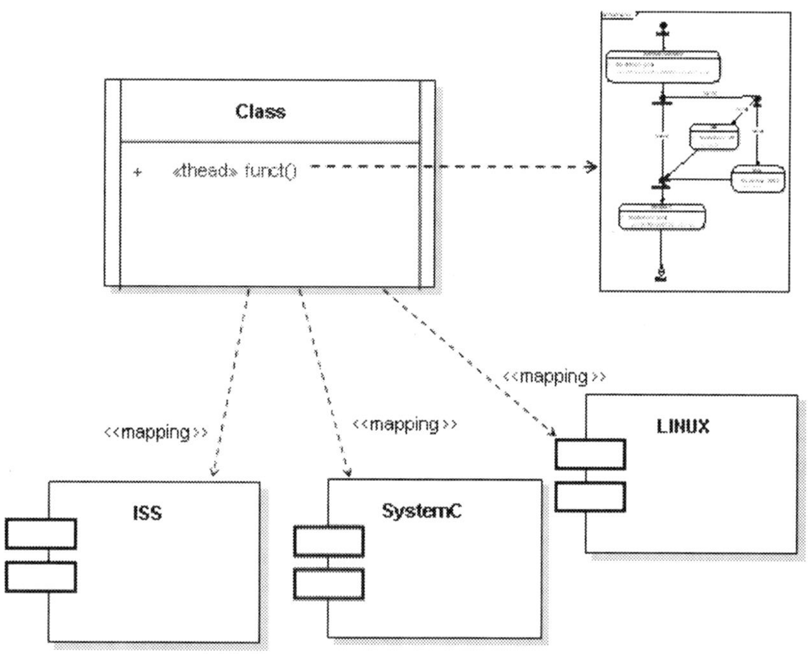

Figure 16.12. Software mapping to implementation.

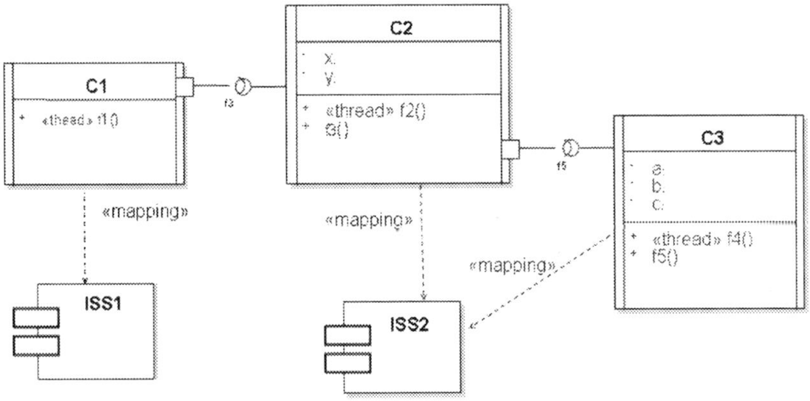

Figure 16.13. Multiprocessor environment.

Communication in MP-environment. In a multiprocessor environment, the mapping relation is used to denote the embedding of parts of the application to the processing elements. In the example in Fig. 16.13, the software is refined down to instruction level by mapping it on two processing elements. In this case, we provide an over simplified view of the hardware architecture just showing the processing elements and their associations; in a more general case it is possible to express also the memory architecture and the memory map.

With a more complex communication it is possible to map the software directly on the host, e.g. on Linux. In the other cases (SystemC and ISS), it is possible to mix the levels of abstraction in the implementation, it is thus possible to map part of the application on SystemC and part on one or more ISSs: in any case a SystemC simulation engine can manage the whole application model by wrapping the needed ISSs.

The three software views differ in the way that the application threads communicate with each other, therefore the refinement between one view and another implies mainly changes in the communication. For example, for the host target view we generate code containing pthreads, as needed by the Linux pthrread.h library, and we rely to the pthreads synchronization mechanism. For example, the pthread_mutex_t could be the data to be associated to the communication and the relative interface includes the pthread_mutex_lock, pthread_mutex_unlock, pthread_mutex_init. On the other hands, if we consider a SW application that runs on a Shared Memory Processor hadware platform, where the API platform follows the OpenMP standard [17], the same mutex communication will require a omp_lock_t data and an interface that includes the function omp_init_lock, omp_destroy_lock, omp_set_lock, omp_unset_lock, and omp_test_lock.

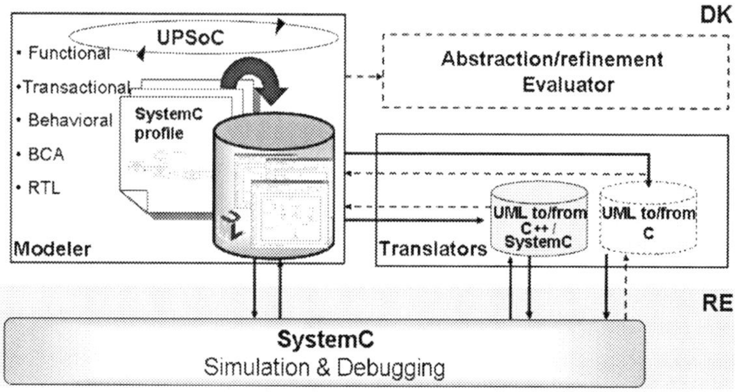

Figure 16.14. Tool architecture.

5. The Codesign Environment

We developed a prototype tool which works as front end for consolidated lower level codesign tools. A more detailed description of this tool is provided in [23]. Figure 16.14 shows the tool architecture. Components visualized inside dashed lines are still under development. The tool consists of two major parts: a development kit (DK) with design and development components, and a runtime environment (RE) represented by the SystemC execution engine. The DK consists of a UML 2.0 *modeler* supporting the UML profile for SystemC [21], *translators* for the forward/reverse engineering to/from C/C++/SystemC, and an *abstraction/refinement evaluator* to guarantee traceability and correctness along the refinement process from the high-level abstract description to the final implementation. This last component is under development.

The modeler. We decided to rely on tools supporting UML 2.0 with the standard extension mechanism of UML profiles. Our current implementation is based on the Enterprise Architect (EA) tool [6] by SparxSystems, but any other tool supporting UML 2.0 can be also used and easily customized. The open API of EA allows us to settle the modeling environment to introduce the SystemC UML profile definition (including the definition of the stereotypes and of the metaclasses they apply to, as well as tagged values, constraints and alternative stereotype images) and use C/C++ and SystemC as action languages.

Figure 16.15 shows a screenshot of EA. The SystemC data types and predefined channels, interfaces, and ports are modeled with the core stereotypes, and are available in the `Project View` with the name `SystemC_Layer1`.

The translators. The EA supports forward/reverse engineering to/from C++. We added to the EA the capability to generate complete C and SystemC

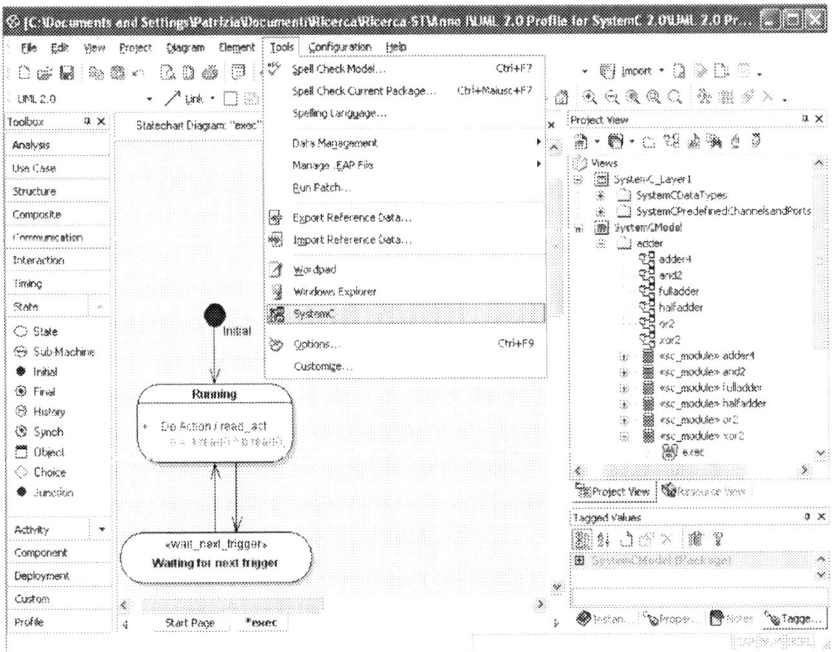

Figure 16.15. Generate SystemC code from EA.

code from UML models for both structural and behavioral aspects. There is, therefore, a unique code generator that makes use of translators to transform parts of the UML model either to SystemC or to C/C++. The generation flow is complemented by a reverse engineering flow that transforms a mixed C/-C++/SystemC program into a UML model. These add-ins were implemented exploiting the automation/scripting interface and the XMI (XML Metadata Interchange) format supported by the EA tool.

6. Case Studies

We have developed several different case studies, some taken from the SystemC distribution like the Simple Bus design, and some of industrial interest. The **Simple Bus** case study is a well-known transactional level example, designed to perform also cycle-accurate simulation. It is made of about 1,200 lines of code that implement a high performance, abstract bus model. The complete code is available at the official SystemC website [18]. We modeled the Simple Bus system in a forward engineering flow. The code generator has been tested primarily on this example.

To test the expressive power of our profile in representing a variety of architectural and behavioral aspects, we modeled the **On Chip Communication**

Network (OCCN) library [16]. The OCCN project [16] focuses on modeling complex on-chip communication networks by providing a highly-parameterized and configurable SystemC library. This library is made of about 14,000 lines of code and implements an abstract communication pattern for connecting multiple processing elements and storage elements on a single chip.

The OCCN design has been imported automatically from the C++/SystemC code into the EA-based modeler exploiting the reverse engineering facility, then it was refined using the modeling constructs of the SystemC UML profile. We have been using this example to test the reverse engineering flow.

In [1] we provide an application example related to a system composed of a VLIW processor developed in ST, called LX, with some dedicated hardware for an **802.11b physical layer transmitter and receiver** described at instruction level. The UML model of this application is a function library encapsulated in a UML class, which provides through ports the I/O interface of the software layer to the hardware system. This class is then translated to C/C++ code and the resulting application code is executed by the LX ISS wrapped in SystemC to allow HW/SW cosimulation at cycle accurate level. The UML wrapper of the LX ISS is modeled with the SystemC UML profile.

7. Related Work

The possibility to use UML 1.x for system design [9] started since 1999, but the general opinion at that time was that UML was not mature enough as a system design language. Nevertheless, significant industrial experiences using UML in a system design process soon started leading to the first results in design methodology, such as the one in [30] that was applied to an internal project for the development of a OFDM Wireless LAN chipset. In this project SystemC was used to provide executable models.

Later more integrated design methodologies were developed. The paper in [19] proposes a methodology using UML for the specification and validation of SoC design. It defines a flow, parallel to the implementation flow, which is focused on high-level specs capture and validation. In [12], a UML profile for a *platform-based* approach to embedded software development is presented. It includes stereotypes to represent platform services and resources that can be assembled together. The authors also present a design methodology supported by a design environment called Metropolis, where a set of UML diagrams (use cases, classes, state machines, activity, and sequence diagrams) can be used to capture the functionality and then refine it by adding models of computation. Another approach to the unification of UML and SoC design is the HASoC (Hardware and Software Objects on Chip) [8] methodology. It is based on the UML-RT profile [25] and on the RUP process [10]. The design process starts with an *uncommitted model* and after a *committed model* is derived by partitioning the system into software and hardware, and then mapped onto a system platform. From these models a SystemC skeleton code can also be generated, but to provide a finer degree of behavioral validation, detailed C++ code must be added by hand to the skeleton code. All the works mentioned above could greatly benefit from the use of new constructs available in the UML 2.0.

SysML [28] is a conservative extension of UML 2.0 for a domain-neutral representation (i.e. a PIM model as in MDA [14]) of *system engineering* applications. It can be involved at the beginning of the design process, in place of the UML, for the requirements, analysis, and functional design workflows. So it is in agreement with our UML profile for SystemC, which can be thought (and effectively made) a customization of SysML rather than UML. Unluckily, no tool support exists for SysML. Similar considerations also apply to the MARTE proposal [26]. The standardization proposal [32] by Fujitsu, in collaboration with IBM and NEC, has evident similarities with our SystemC UML profile, like the choice of SystemC as a target implementation language. However, their profile does not provide building blocks for behavior modeling and does not adopt any time model.

Some other proposals already exist about extensions of UML toward C/C++/-SystemC. All have in common the use of UML stereotypes for SystemC constructs, but not rely on a UML profile definition. In this sense, it is appreciable the work in [2] attempting to define a UML profile for SystemC; but, as all the other proposals, it is based on the previous version of UML, UML 1.4. Moreover, in all the proposals we have seen, no code generation, except in [15], from behavioral diagrams is considered.

8. Conclusions and Future Work

The work presented here is part of our ongoing effort to enact design flows that start with system descriptions using UML-notations and produce full implementations of the SW and HW components as well as their communication interfaces. Currently, we are defining a unified process called UPES (Unified Process for Embedded Systems) which is intended to assist the designers across the UML modeling activity from a high-level functional model of the system down to a RTL model by supporting current industry best practice in *platform-based design* [33, 12]. The MDA [14] principles are aimed at providing executable models and at supporting automation of the primary UPES activities.

We are still exploring the possibility to implement MDA-style transformations for both the HW and the SW descriptions to allow a PIM to be transformed into a PSM, and an abstract PSM to be transformed into a refined PSM according to the levels of abstraction: functional, transactional, behavioral, BCA, and RTL. In particular, we are defining a formal refinement methodology with precise abstraction/refinement rules for the *transactional-level modeling*. Transactional models are used for functional modeling of the communication enhanced with actual timing information. To achieve this goal, we have been working to a revision of the SystemC UML profile to include the **OSCI TLM 1.0 library** [18] and the new features provided by **SystemC 2.1** (like `sc_export` ports, the `event_queue` mechanism, dynamic processes, fork/join synchronization) for modeling at the TLM level of abstraction.

References

[1] S. Bocchio, E. Riccobene, A. Rosti, and P. Scandurra. A SoC design flow based on UML 2.0 and SystemC. In: *DAC Workshop UML–SoC'05.*

[2] F. Bruschi and D. Sciuto. SystemC based design flow starting from UML Model. In: *Proc. of European SystemC Users Group Meeting*, 2002.

[3] J. Bézivin. In search of a basic principle for model driven engineering. *CEPIS, UPGRADE, The European Journal for the Informatics Professional*, V(2):21–24, 2004.

[4] J. Bézivin. On the unification power of models. *Software and System Modeling (SoSym)*, 4(2):171–188, 2005.

[5] W. Cesfirio, A. Baghdadi, L. Gauthier, D. Lyonnard, G. Nicolescu, Y. Paviot, S. Yoo, A. Jerraya, and M. Diaz-Nava. Component-based Design Approach for Multicore SoCs, 2002.

[6] Enterprise Architect: www.sparxsystems.com.au/.

[7] ECSI Institute Workshop: UML Profiles for Embedded Systems, March 27–28, 2006. www.ecsi-association.org/ecsi/.

[8] M. Edwards and P. Green. UML for Hardware and Software Object Modeling. *UML for Real Design of Embedded Real-Time Systems*, pp. 127–147, 2003.

[9] G.Martin. *UML and VCC.* White paper, Cadence Design Systems, Inc, Dec. 1999.

[10] P. Kruchten. *The Rational Unified Process.* Addison-Wesley, 1999.

[11] L. Lavagno, G. Martin, and B. Selic. *UML for Real Design of Embedded Real-Time Systems.* Kluwer, 2003.

[12] L. Lavagno, G. Martin, A. Sangiovanni Vincentelli, J. Rabaey, R. Chen, and M. Sgroi. UML and platform-based design. *UML for Real Design of Embedded Real-Time Systems*, 2003.

[13] G. Martin and W. Mueller. *UML for SoC Design.* ISBN 0-387-25744-6, Springer, July 2005.

[14] OMG. The Model Driven Architecture (MDA). http://www.omg.org/mda/.

[15] K.D. Nguyen, Z. Sun, P.S. Thiagarajan, and W.F. Wong. Model-Driven SoC Design: The UML-SystemC Bridge. *UML for SOC Design*, 2005.

[16] OCCN Project: http://occn.sourceforge.net/.

[17] The OpenMP standard. http://www.openmp.org.

[18] The Open SystemC Initiative. http://www.systemc.org.

[19] Q.Zhu, R.Oishi, T.Hasegawa, T.Nakata. System-on-Chip Validation using UML and CWL. In *Proc. of CODES*, 2004.

[20] E. Riccobene and P. Scandurra. Modelling systemC process behaviour by the UML method state machines. In: *Proc. of Rapid Integration of Software Engineering Techniques (RISE'04)*. Springer, LNCS 3475.

[21] E. Riccobene, P. Scandurra, A. Rosti, and S. Bocchio. A UML 2.0 Profile for SystemC. ST Microelectronics Tech. Rep. AST-AGR-2005-3.

[22] E. Riccobene, P. Scandurra, A. Rosti, and S. Bocchio. A SoC design methodology based on a UML 2.0 profile for SystemC. In: *Proc. of Design Automation and Test in Europe*. IEEE, 2005.

[23] E. Riccobene, P. Scandurra, A. Rosti, and S. Bocchio. A model-driven design environment for embedded systems. In: *DAC '06: Proc. of the 43rd Annual Conference on Design Automation*, pp. 915–918, New York, 2006. ACM Press.

[24] Patrizia Scandurra. Model-driven Language Definition: Metamodelling Methodologies and Applications. Ph.D thesis, University, of Catania, Italy, December, 2005.

[25] B. Selic. A generic framework for modeling resources with UML. In: *Proc. of the 16th Symposium on Integrated Circuits and Systems Design (SBCCI'03)*, 33:64–69, 2000, IEEE.

[26] R. De Simone et al. MARTE: A New Profile RFP for the modelling and analysis of real-time embedded systems. In: *DAC Workshop UML-SoC'05*.

[27] OMG. UML profile for schedulability, performance, and time, formal/03-09-01.

[28] SysML Partners website: http://www.sysml.org/.

[29] T. Gröetker, S. Liao, G. Martin, and S. Swan. *System Design with SystemC*. Kluwer Academic Publishers Norwell, MA, 2002.

[30] T. Moore, Y. Vanderperren, G. Sonck, P. van Oostende, M. Pauwels, and W. Dehaene. A design methodology for the development of a complex system-on-chip using UML and executable system models. In: *Proc. of FDL'02*.

[31] OMG. UML 2.0 Superstructure Final Adopted Specification, ptc/04-10-02.

[32] Fujitsu Limited, IBM, NEC. A UML Extension for SoC. Draft RFC toOMG, 2005-01-01.

[33] A. Sangiovanni Vincentelli. Defining platform-based design. *EEDesign of EETimes*, February 2002.

Chapter 17

A METHOD FOR MOBILE TERMINAL PLATFORM ARCHITECTURE DEVELOPMENT

Klaus Kronlöf, Samu Kontinen, Ian Oliver, and Timo Eriksson
Nokia Research Center
Itämerenkattu 11-13
00180 Helsinki, Finland

Abstract We introduce a novel architecture, called the Network-on-Terminal Architecture (NoTA), for mobile terminal platforms. This paper concentrates on the platform development and validation flow adopted for NoTA. Platform requirements are expressed as use cases that are modelled using UML2 with Telelogic's Tau G2 tool. Models are executable so that use case behaviour can be animated. Use cases are used as test cases in the platform architecture development for which use case information is transferred as execution traces. We use CoFluent Studio tool for platform architecture specification and performance analysis. The use case execution trace is fed into a functional model that represents the computation load. NoTA is service oriented and thus the functional model consists of platform services. The computation and communication resources are modelled with a separate platform architecture model. The tool allows exploring different configurations and allocations of the functional and platform models quickly and provides extensive performance information, including power consumption.

Keywords Platform architecture, service-oriented architecture, model-based engineering, web services, UML, MCSE.

1. Motivation

Digital convergence and mobile device industry horizontalisation are creating pressure for companies to renew their competences as well as the device architecture. Current CPU centric highly integrated one-for-all-platforms have come to the end of the road. Future mobile device architectures are system-wise

S.A. Huss (ed.), Advances in Design and Specification Languages for Embedded Systems – Selected Contributions from FDL'06, 285–300.

modular and service based. Network On Terminal Architecture (NoTA) is such an architecture.

The development of mobile terminal platforms should start from end-user needs. In our company we have traditionally expressed them as use cases. For NoTA we have developed a more rigorous model-based method of presenting use cases and using them to guide platform architecture development. The method also utilises the service oriented nature of NoTA to form an intermediate functional model between abstract use cases and the platform architecture solution.

In the method we use commercial tools and standard modelling languages as much as possible. The innovation is in integrating them to support use case driven development of service oriented platform architectures.

2. Introduction to Nota

NoTA is an interconnect centric modular service-oriented architecture for today's and future mobile device platforms. NoTA claims to provide superior performance and to make effective horizontalisation possible via eased integration. The development method associated with NoTA ensures that designs are stepwise verifiable against end-user requirements. The method is also flexible and scaleable with reuse on different levels. NoTA allows the use of novel technologies and open innovation, and shortens the R&D cycle.

A NoTA platform consists of loosely connected services running on top of heterogeneous subsystems. In NoTA based systems all service and data communication is routed via the network stack as shown in Fig. 17.1; this approach is similar to that taken formerly by CORBA [7] and lately in a more sophisticated

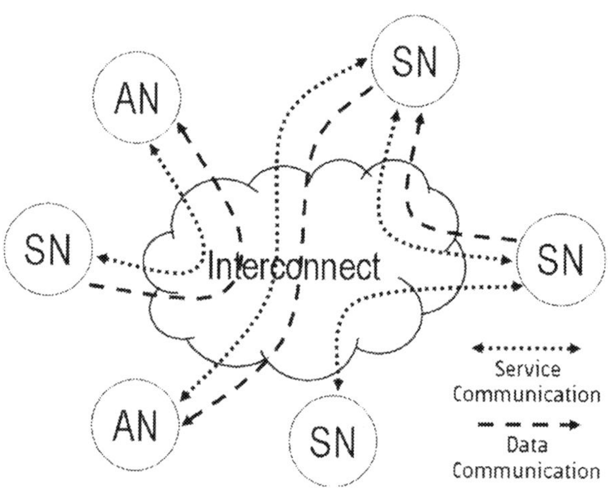

Figure 17.1. NoTA logical architecture consisting of three types of foundation elements called ApplicationNodes, ServiceNodes, and Interconnect.

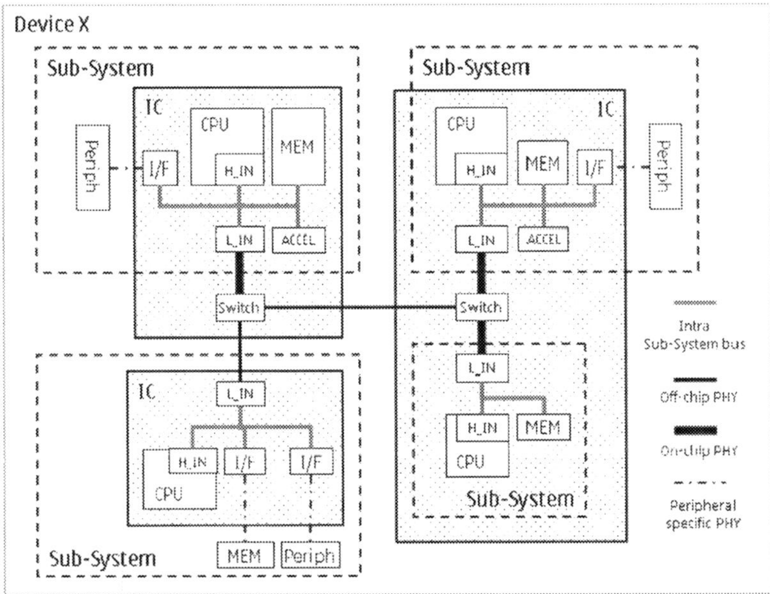

Figure 17.2. Example of physical implementation of NoTA based platform.

form by web services [8]. NoTA takes these principles and specialises them for use in a highly embedded system. The NoTA method includes a platform development flow that ensures that services, subsystems and the interconnect topology are matched to end-user requirements. It also provides formal reusable specifications for the platform entities.

NoTA defines two main level of protocols for the interconnect, called H_IN and L_IN. H_IN is a high level protocol stack providing communication functionality for platform services and applications. L_IN is the low level protocol that provides the physical connection between subsystems.

A NoTA subsystem implements a set of services. It is an architectural concept that does not necessarily align with chip boundaries. So, we may have several subsystems on a chip and a subsystem may extend outside the boundaries of a chip, as illustrated in Fig. 17.2.

3. Nota Platform Architecture Development Method

The industrial practice in platform architecture development is quite informal and heavily relies on system architect's experience. This is feasible when changes in successive generations of the architecture are relatively small, but is problematic when dealing with truly novel architectural concepts that call for systematic exploration of quite different alternatives. Furthermore, platform requirements are typically expressed in technical terms that are not properly connected to end-user needs. The NoTA platform architecture development method aims at overcoming these pitfalls of industrial practice.

NoTA-based systems are engineered in a systematic requirements driven manner. It is characterised by the following principles.

Separation of concerns. We want to be able to develop different aspects of the system independently from each other in order to manage complexity and to facilitate reuse. In the NoTA method we separate the domains of:

- End-user requirements
- Platform functionality, i.e. services provided by the platform
- Platform architecture, i.e. definition of subsystems and communication infrastructure [5]
- Implementation of subsystems (SW and HW) and interconnect protocols (SW and HW)

Each of the domains has their own self-contained models. Eventually, in the final system, these domains are of course related to each other, but we want to be able to postpone fixing these relations until the time we actually define the system instance (that can be a product or a product platform).

Model-based engineering. In the NoTA method the artefacts developed in different phases of the process are models with well-defined semantics. We want to avoid misunderstandings and the consequent errors caused by ambiguousness and hidden meanings of informal documentation. We also want to be able to use analysis, verification, transformation, code generation, and synthesis tools that operate on models.

Reuse of models. We believe that the possibility to effectively reuse models in different contexts gives a big improvement of design productivity compared to conventional methodologies. In the NoTA method different kinds of models are stored in repositories from where they can be retrieved and used to compose new system configurations. We have put special emphasis on modelling techniques that enable easy composition of models.

Early validation and verification. One motivation of model-based engineering is early validation and verification of specifications and designs. In the NoTA method the validation and verification [13] start already at end-user requirements phase with executable use case models. Later on, its focus is on the correctness of platform specification and performance analysis at both specification and implementation phases. In the NoTA method, the validation and verification is not limited to logical correctness, but covers also non-functional aspects, such as real-time performance and energy consumption.

4. Requirements Modelling

In NoTA we have taken a use case driven approach to requirements modelling. Generally speaking, a use case captures a contract between the stakeholders of a system about its behaviour [1]. It describes the system's behaviour under various conditions as it responds to a request from one of its stakeholders, called the primary actor. The primary actor initiates an interaction with the system to accomplish some goal.

The classical use case approaches [1] concentrate on specifying system functionality by means of action sequences. We have developed these approaches further because we feel that it is essential to be able to express concurrency in functional requirements and relationships between different requirements aspects (functional, non-functional, user interface, and interoperability). The process of capturing and modelling requirements is depicted in Fig. 17.3.

Classical use case descriptions are textual. We believe that in some cases graphical forms, such as sequence diagrams, are a more natural form to describe a use case. We see that textual and graphical forms are just two alternative presentations for the same thing. Ideally the tool environment should support different presentation forms of a coherent underlying use case model.

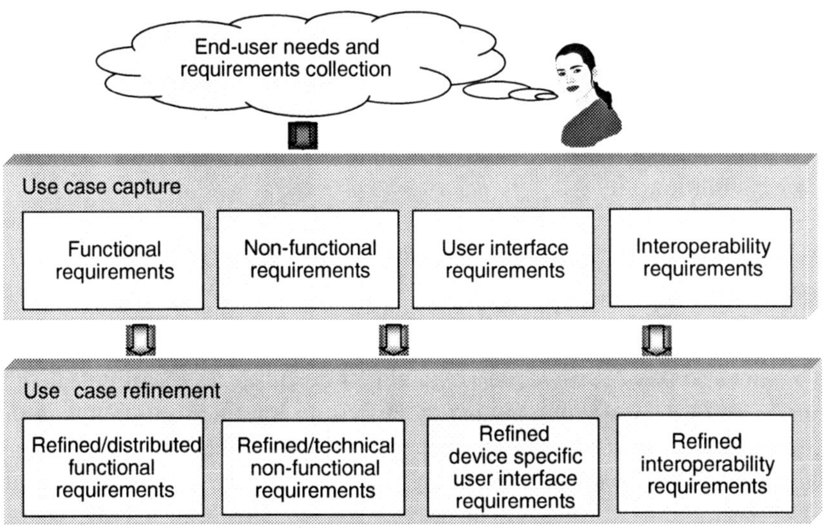

Figure 17.3. Requirements capture and modelling process.

Functional Requirements

Classical use case approaches are quite developed in the functional aspects, so we adopt these concepts directly. Basically we define different kinds of actors and the actions they take in the use case.

In the textual form we first define the actors (primary actor, other external actors, subject), and then describe the actions and their ordering in the main success scenario [1]. Actions are written in active form so that the (initiating) actor that takes the action, and the possible receiving actors, are explicitly named.

In the graphical sequence diagram, form actors are represented by vertical lines and actions are represented by horizontal arrows that extend from the initiating actor to the receiving actor. The ordering of actions is from top to bottom. More complex ordering (concurrency, alternative) can be expressed with special notations.

Hierarchy is a useful way to hide details in a complex use case. Hierarchy can be used in both textual and graphical forms. In the action dimension we can define composite actions that represent a sequence (or other ordering) of lower level actions. In the actor dimension we can define composite actors in basically the same way. However, representing actor hierarchy in textual form is non-trivial (while in graphical form it is straightforward) and, therefore, we limit the hierarchy in our method to the action dimension. Note that composite actions can be applied recursively to create a hierarchy of arbitrary depth.

Classically the actions follow each other sequentially in the main success scenario. This is fine for simple use cases, but sequential definition becomes very complicated or almost impossible when we need to combine loosely coupled (almost independent and concurrent) functionalities in the same use case. We feel that it is very important to allow concurrency in order to avoid over-constraining the use case. After all, ordering of the steps is an additional requirement that has to be explicitly justified, because it limits the implementation possibilities.

Non-functional Requirements and User Interface Requirements

Most non-functional requirements concern a specific part of the use cases (e.g. response time to a user request or quality of picture at a specific point). This is true for real-time requirements as well as for user interface requirements and, for example, security requirements. We want to attach this kind of requirements explicitly to the relevant actions of the use case instead of specifying them separately from the use case. On the other hand, we also want to have a purely functional view of the use case available, since the amount of information in the use case tends to grow so big that it obscures the essential function (hierarchical definition of actions is another way to hide details).

Ideally the tool would allow adding non-functional and user interface requirements to either textual or graphical presentation of the main success scenario. For example, it is convenient to represent a response time as a line segment in the sequence diagram. Furthermore, the tool should support multiple views of the use case for different purposes. All these should be based on a coherent model so that the consistency of different views is guaranteed.

General End-user Requirements and Business Requirements

It is difficult to associate to a single use case. A typical such requirement refers to the size/weight of the device (e.g. pocket-size), robustness (e.g. waterproof), battery life (although this is actually dependent on the use cases), etc. In use case descriptions non-functional requirements are primarily intended for requirements that are specific to that use case (or part of it). Normally we do not include general requirements in the use case description.

Some technology and component choices depend almost exclusively on business environment and have very little to do with the actual end-user requirements. Business requirements refer to legacy technologies, partners, other leading business players, etc. We omit this kind of requirements completely in our method, although we understand that it can be very important from business perspective.

Executable Use Case Models

We use UML2 [9] as the syntax of executable use case models for the purposes of clarification or demonstration that is often necessary in the negotiations with the stakeholders. The semantics of executability is provided by the SDL [2] profile of the UML2 embedded in the chosen toolset, i.e. Telelogic's Tau G2.

The ability to execute a use case makes it more demonstratable to the customer and allows the modeller to obtain early feedback regarding how that particular piece of functionality works and how it interacts with other aspects of the system in question [10, 11]. One disadvantage here is that use cases are specified in an imperative way thus guiding the modeller to a certain implementation path; this is opposite to approaches based upon declarative specification which makes no or very little architectural choices. However, imperative modelling is more common and the modeller easily educated in how to avoid implicit architectural decisions [12].

In our approach we create a separate model for each of the actors (we use the term "actor" here in the same way as in [1]) of the use case, and the actions themselves are modelled as interactions between these models. In UML2, each actor is modelled as an active class associated with a state machine diagram.

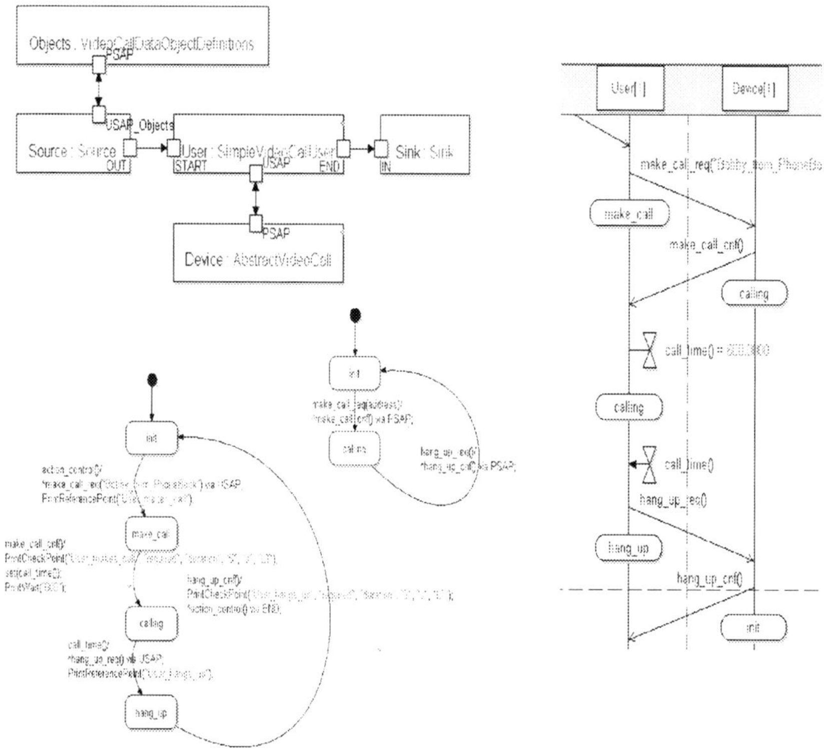

Figure 17.4. An executable UML2 model of a simple use case.

A composite structure diagram [3] expresses configuration of the use case. Interactions are modelled as message (or signal) interchanges (asynchronous communication). A simple example is shown in Fig. 17.4.

In a simple use case each actor is modelled with a single state machine. Complex use cases are often composed of simpler ones so that the model of each actor is a composition of simpler actors' models. The primary actor (normally the user) is composed in a way that expresses the sequential, concurrent, or alternative composition of simpler behaviours. The semantics is analogous to the corresponding composite actions of the textual use cases. Fig. 17.5 shows an example of sequential composition.

In most cases the communication between the actors in the use case is essentially synchronous in the sense that the initiating actor normally waits for some kind of response before continuing its behaviour (e.g. the user wants to see something happening on the display of the device). We model synchronous communication with two-way asynchronous communication, that is, with a request-confirmation message/signal pair.

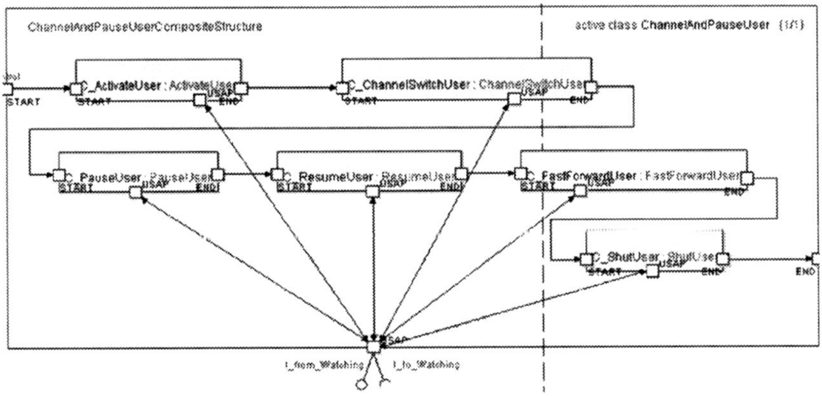

Figure 17.5. Example of a composite primary actor with sequential composition.

Non-functional requirements are treated as attributes of specific elements of the model. For example, real-time requirements are refined as (allowed) state transition durations and communication latencies.

As mentioned in the previous chapter, in a practical business situation there are non-functional and user interface requirements that have nothing to do with the end-user needs or even with the use case. Our method does not cover such requirements at the moment.

Abstract and Refined Use Case Models

An abstract use case treats all actors of the use case as black boxes. We are only interested in the externally observable behaviour of the actors. Furthermore, we consider only high level actions that are meaningful to the end-user to accomplish her or his goal.

Use case refinement splits the end-user actions into smaller technical/internal actions of the target system. The goal is to eventually map all the actions to predefined NoTA services of the execution platform. Refinement is typically done in many steps. The number of step depends on the size and complexity of the use case. The nature of the target services (primitive versus. high level) also affects the number of refinement steps needed. Each refinement step is verified, which means ensuring that the functional behaviour is preserved as well as that the non-functional constraints are satisfied. The particular technique used for verification also affects the number of refinement steps needed. Currently, refinement here means that the behaviour of the system remains constant through decomposition rather than the stricter mathematical notion [14] and similar [15].

The refinement steps concern the subject actor only; all the other actor models remain the same. In these refinement steps the state machine model of the subject is split hierarchically and recursively into multiple state machines that communicate with each other. In each refinement step we must ensure that the externally observable behaviour (i.e. the message communication at external interfaces) of the subject remains the same.

The goal of the refinement is to finally express the use case subject's behaviour using predefined NoTA services. Each predefined service is also modelled as a state machine with a communication interface. The fully refined subject model consists of a hierarchical structure where the leaves (lowest level objects) are predefined services (with predefined state machine models). Fig. 17.6 shows an example of a refined subject.

A use case often involves some kind of end-to-end system behaviour that may extend outside the borders of the target system (e.g. a single terminal). We have to accommodate to this in our method, while at the same time translate the required behaviour into requirements for the terminal system. For example, we need to decide what part of the functionality goes into the terminal or how to divide an end-to-end performance requirement between the devices on the network. These high level architectural design decisions are done in the first refinement step. The second refinement step splits further the functionality allocated to the terminal and maps it to predefined services.

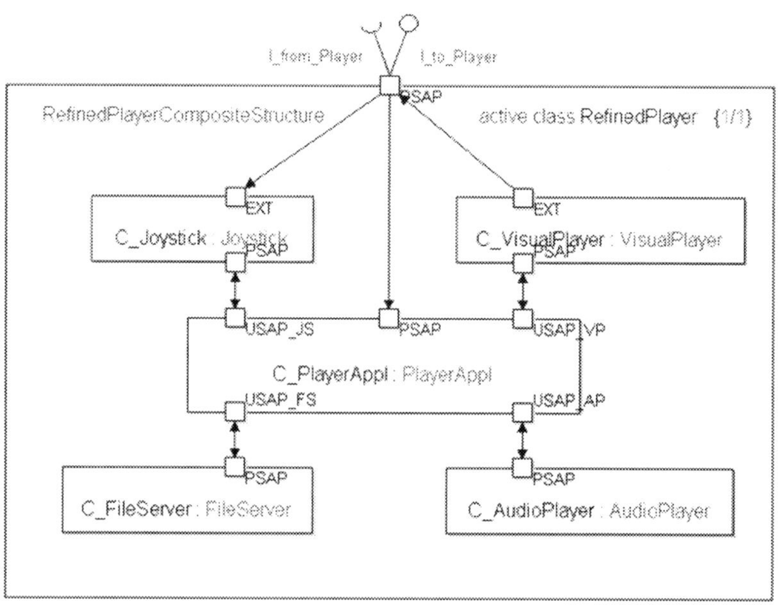

Figure 17.6. A composite structure diagram of a refined subject.

5. Architectural Modelling

We have adopted the MCSE method [4] for architectural modelling in NoTA. According to it, an architectural model is developed by building the functional architecture (timed behavioural model, e.g. functional model of the system with timing information) and the platform architecture (executive structure) and mapping the functional blocks onto the executive structure. The CoFluent studio toolset [6] includes tools that support model creation and mapping according to MCSE.

The requirements for a NoTA based platform come from the end-user requirements expressed as use case models as described above. The selected collection of use cases is first studied and all services used in the refined use case models (called primary services) are identified. Next, the set of required services is reduced to minimize overlap and redundant services. Also, at this point, if there are several versions of the same service needed, the version that fulfils all the needs is selected and others discarded. As a result of this process, the set of required primary services is defined.

The use case models together with the set of required primary services are used to build the functional architecture model. It consists of Service Node (SN) models and Application Node (AN) models. A SN model represents an instance of a service (there may be several instances of the same service) and an AN model defines the way the application uses the services in a particular use case.

In NoTA, a service is specified in a special format called Service Interface Specification (SIS). SIS includes the interface signature of the service in question and a description of its externally observable behaviour expressed as a Finite State Machine (FSM). SIS also includes the relevant non-functional attributes of the service, such as timing and power consumption. In architectural modelling, SN models are derived directly from the SIS. AN models are derived from the execution traces of use case models.

The platform architecture model consists of blocks representing the subsystems and routing switches. Mapping the SNs and ANs into the subsystems and defining the communication network topology among the subsystems yields the architectural model. The Interconnect Node (IN) functionality is integrated into the components of the platform architecture model.

In the following subsections and in Fig. 17.7 we explain how the MCSE method is applied to NoTA using CoFluent.

Functional Architecture

The functional or timed behavioural model describes the logical partitioning and behaviour of the system. As said before, in NoTA the functional model consists of Service Nodes (SN) and an Application Node (AN). The SNs include

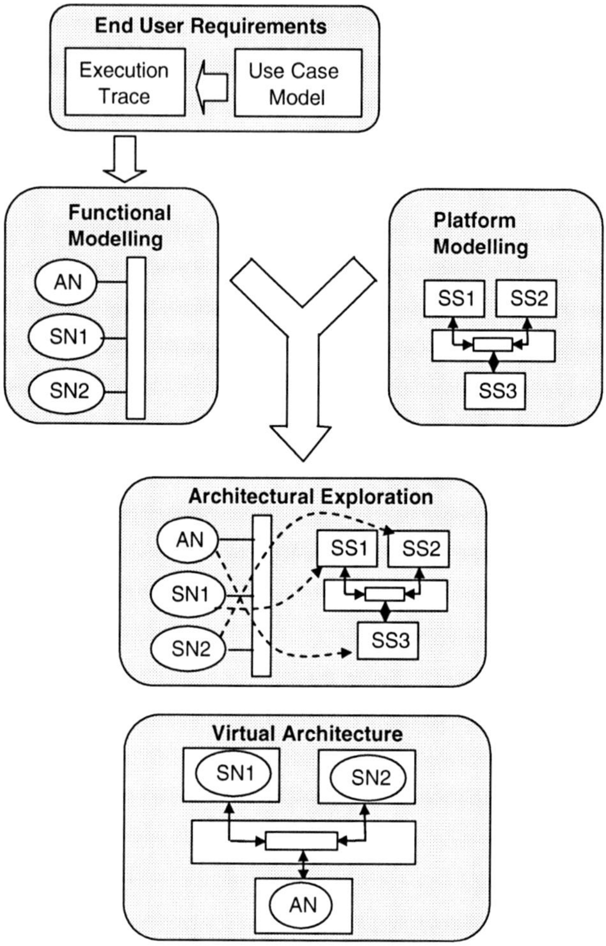

Figure 17.7. The MCSE method is applied to NoTA. CoFluent Studio is used as the platform modelling tool.

all primary services of the use case and any additional secondary services used by the primary services.

The functional editing tool of CoFluent Studio captures the graphical description of the SNs and AN. The internal model of each SN is derived directly from the corresponding SIS. The behaviour of the AN is defined by the use case and we use the generated XML trace here. The service requests to SNs are modelled as messages. In NoTA all service requests are passed over the IN. The functional behaviour of a system consisting of AN, SNs and ideal IN can be simulated independently without a definite platform architecture.

In NoTA, the behaviour of SNs are modelled as finite state machines. As parts of a CoFluent model they must be represented either as SystemC models or as functional structures (CoFluent functions), which are lower hierarchical level models used as "black boxes" within the model. The behaviour of the CoFluent functions can be further defined with algorithms written in C or C++. We have adopted the latter approach with additional C++ algorithms to read in and interpret the XML files. A parser module extracts the needed information from the XML-model and assigns them to the correct placeholders in the CoFluent SN template.

As mentioned earlier, there are two main types of communication in the NoTA network. Models for the data object communication is added to the SNs to get insight about the data amount between different nodes. In practice the same functional link is used for both the service communication and the data traffic. The type of the link and the message that is sent into it are modelled as a C++ class. It contains fields for routing, message type and size and sub classes for the content.

The use case behaviour is imported to the CoFluent model as a XML-trace file. The file consists of a sequence of service requests with possible additional parameters. In CoFluent the file is read in the AN and corresponding service requests are sent to the correct services.

Platform Architecture

The platform model is an abstract representation of the physical architecture. CoFluent provides a set of building blocks for platforms: processors, shared memories, signals, and connections that use communication nodes. At this design step the subsystems are outlined as placeholders for the SNs and the AN. One subsystem consists of one processor, where the SNs are run in parallel. The actual implementation of the subsystem is not modelled. The Routing Switch (RS) is modelled with a routing model of CoFluent which contains built-in performance statistics gathering functionalities. One processor is reserved for each RS and the subsystems are connected to the RS with communication nodes. The resulting network topology represents the accurate interconnect that is needed in the architectural simulations.

It should be noted that the SN and AN models contain performance data (e.g. time to process a service call). Hence, the CoFluent processor type should be selected as hardware processor capable of running the SN models independently of each other. Later on when running real SW models for certain SNs, the SW processor models become handy.

The platform models themselves have tuneable parameters that affect the overall system performance. For example, the bandwidths of the RS's can be configured independently.

Architectural Exploration

The architecture design consists of three main parts, namely decision about the number (and type) of subsystems in the device, Interconnect topology between the subsystems and finally mapping of the SNs and ANs into the subsystems.

A subsystem can be defined as a collection of SNs that has an IN connection. One subsystem can contain several SNs. For example: a storage subsystem can act as a conventional mass storage or as a streaming media server. These require completely different services but as they use the same hardware resources it is beneficial to locate them inside the same sub-system. The SNs can be distributed across the sub-systems in several different ways and accordingly several architectural configurations can be evaluated to identify the potential bottlenecks and to maximize the system performance.

Network traffic analysis is a key output in verifying the designed architecture. The way to present the analysis results is one of the topics to be defined later on. Each designed architecture needs to be simulated against all the use cases. There are certain parameters related to the SNs, ANs, INs and RSs and that could be optimised during the architecture design. Local buffer sizes are maybe the most important of such parameters.

Steps within the architectural exploration:

1. Define the number and types of subsystems
2. Define the interconnect topology to connect the subsystems (number and types of routing switches to be decided here)
3. Map SNs and ANs into the subsystems
4. Run simulations against the original use cases
5. Analyse results
6. Go back to step 3 to optimise the current architecture
7. Go back to step 1 to try different architectures
8. Choose the most optimal case(s)
9. Virtual architecture

After exploration of different architectures and decision on the most optimal case, the architecture solution consists of the following:

- A set of sub-systems connected together with a certain interconnect topology
- Mapping of the decomposed use case originated services into the above subsystems
- Verified and refined performance parameters for the services
- All of the above verified against the original end-user use cases
- These set the requirement specifications for each subsystem.

6. Discussion

We have presented our method for developing service oriented platform architectures for mobile terminals starting from end-user requirements. In many cases the choice of a particular technique depends on the product development culture in the company and also on the constraints of available tools.

We express end-user requirements as use case because it is the company practice. However, we wanted to add rigor to the approach. We wanted executable models that can be animated, because this facilitates validation with the stakeholders. The decision to use UML 2.0 for modelling was quite obvious since it is becoming the system level modelling standard in the company. Telelogic's Tau G2 tool was available at the time, so we took it.

The characteristics of the tool have sometimes unwanted consequences in the method. The only practical way to create executable models in Tau G2 is through state machines. This is a quite operational and implementation oriented way. UML 2.0 offers activity diagrams and sequence diagrams that might suit better here. Some non-UML methods, such as CSPs, might be even better. In principle we would prefer declarative methods but at the same time we want the model to be executable.

The choice of CoFluent Studio for architectural exploration was also straightforward for many reasons, including company practices, ongoing cooperation, and certain excellent capabilities of the tool. However, the tool is not based on UML. The link between UML-based requirements modelling and non-UML architecture exploration is realised as execution traces. This is a rather weak link. It is a one-directional link, so it does not allow true model-based testing.

Ideally we would like to use the use case model directly, so that we could repeat the animation at any point of development and see the impact of design decisions. This could be achieved through deep tool integration based on a common meta-model shared by all tools. The integration task would be easier if all tools were UML-based, so the meta-model could be specified as a UML profile.

References

[1] Cockburn, A. and *Writing Effective Use Cases*. Addison-Wesley, 2000.

[2] Ellsberger, J., Hogrefe, D., and Sarma, A. *Formal Object-oriented Language for Communicating Systems*. Prentice-Hall, 13-632886-5, 1997.

[3] Bock, C.: UML 2 Composition Model. *Journal of Object Technology*, 3(10), 2004.

[4] Perrier, V. System architecting complex designs, *EmbeddedSystems Europe*, 8(55), Feb 2004.

[5] Len Bass, P. C., Kazman, R., *Software Architecture in Practice*, 2nd edn. Addison-Wesley, 0-321-15495-9, 2003.

[6] CoFluent Design home page, http://www.cofluentdesign.com

[7] Object Management Group, The Common Object Request Broker Architecture. 1995. http://www.corba.org/

[8] Erl, T. *Service-Oriented Architecture: Concepts, Technology, and Design*. Prentice-Hall, 0131858580, 2005.

[9] UML 2.0 Superstructure http://www.omg.org/technology/documents/formal/ uml.htm

[10] Fuchs, N. E. Specifications Are (Preferably) Executable. *IEE/BCS Software Engineering Journal*, 1992.

[11] Bourhfir, C. Dssouli, R. Aboulhamid, E. and Rico, N. Automatic Executable Test Case Generation for Extended Finite State Machine protocols. In: IWTCS'97 [17], pp. 75–90.

[12] Oliver, I. Demonstration of the Proof of Concept for Specifying and Synthesizing Hardware using B and Bluespec. FDL06.

[13] Oliver, I. Experiences in Using B and UML in Industrial Development. In: J. Julliant and O. Kouchnarenko (eds): *B2007, Lecture Notes in Computer Science 4355*, pp. 248–251. Springer, Berlin, 2006.

[14] Back, R.-J. and Wright, J. Refinement Calculus: A Systematic Introduction, (Texts in Computer Science) 0387984178, 1998.

[15] Milner, R. *Communicating and Mobile Systems: The ?-Calculus*. Cambridge University Press, 0-521-65869-1, 1999.

Chapter 18

UML2 PROFILE FOR MODELING CONTROLLED DATA PARALLEL APPLICATIONS

Ouassila Labbani[1], Jean-Luc Dekeyser[1], Pierre Boulet[1], and Éric Rutten[2]

[1]*LIFL - UMR USTL/CNRS 8022*
Bâtiment M3, Cité scientifique
59655 Villeneuve d'Ascq Cedex FRANCE
{labbani; dekeyser; boulet}@lifl.fr

[2]*INRIA Rhône-Alpes*
ZIRST, 655 avenue de l'Europe, Montbonnot
38334 St Ismier Cedex FRANCE
rutten@inrialpes.fr

Abstract In this paper, we present a UML2 profile introducing control in the Gaspard2 data parallel applications using the synchronous approach. This concept allows to take the change of running mode into account in the case of parallel applications. It is then possible to study more general systems mixing control and data parallel processing.

Keywords Data parallel applications, reactive models, UML profile

1. Introduction

Computation intensive multidimensional applications are more and more present in several application domains such as image and video processing or detection systems. The main characteristics of these applications are that they perform intensive computations, they operate in real-time conditions and are generally complex and critical. They are also multidimensional since they manipulate multidimensional data structured into arrays, and their data can generally be executed in parallel.

S.A. Huss (ed.), Advances in Design and Specification Languages for Embedded Systems – Selected Contributions from FDL'06, 301–317.

In this paper, we are interested in modeling embedded applications mixing control and data parallel processing. We propose a UML profile for these systems which allows and facilitates their high-level study. Our proposition is applied to the Gaspard2[1] environment and studies the control introduction in Gaspard2 models using the synchronous approach.

2. Data Parallel Applications Modeling

Some computation models and design methodologies have been proposed to study and model parallel applications. In this paper, we will focus on the high level modeling of this kind of applications by using the Gaspard2 UML profile.

Gaspard2 Environment

Gaspard2 is an Integrated Development Environment for SoC (System on Chip) visual comodeling. It extends the Array-OL model [1] and allows modeling, simulation, and code generation of SoC applications and hardware architectures. The Gaspard2 environment is mainly dedicated to the specification of signal processing applications. It is based on a *model oriented* methodology according to a Y design flow. Concepts of each design level in this model are represented independently of the execution or simulation platforms.

The Y model in Gaspard2 is mainly defined around three metamodels: the *application* which allows to specify the functionality of the system, the *hardware architecture* which will be used to model the hardware platform supporting the execution of the application and to perform its functionalities, and the *association* which allows to specify the mapping of the application on a given hardware architecture. In this approach, the concepts and semantics of each model are abstract since no component is associated with an execution, simulation, or synthesis technology.

The starting point in Gaspard2 consists in modeling the application, the hardware architecture, and the association by using the Gaspard2 UML profile [2, 5]. These models are then imported and studied by applying mapping and scheduling algorithms and automatic SystemC code generation. The model definitions of the Gaspard2 environment are based on a *component oriented* methodology. This methodology makes it possible to clearly separate the different parts of the Y model, and facilitates the reuse of existing software and hardware IPs.

Gaspard2 UML Profile

To model the different Gaspard2 concepts, a first UML2 profile version has been proposed [2, 5]. In that version, the profile is defined around five packages: `component`, `factorization`, `application`, `hardware-Architecture`, and `association` as shown by Fig. 18.1.

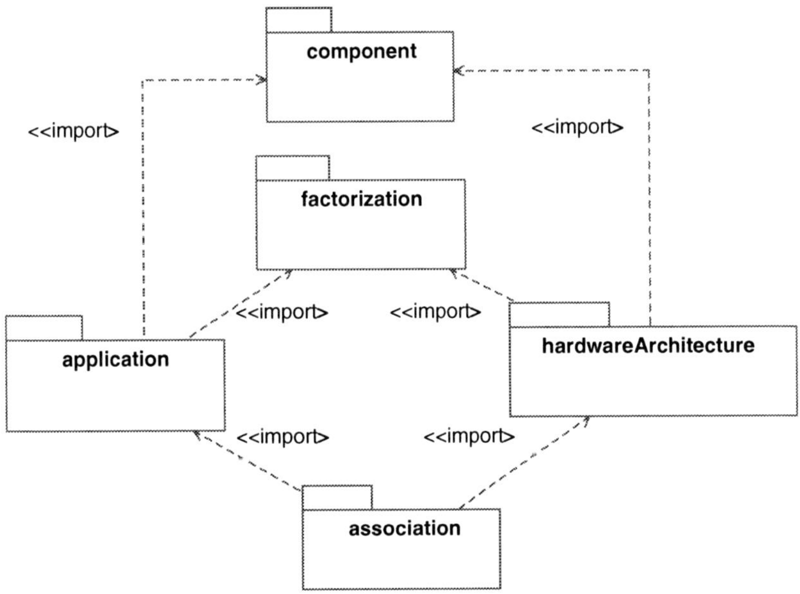

Figure 18.1. Different packages of the Gaspard2 UML profile.

The `application` and the `hardwareArchitecture` packages specify respectively the software application of the system and the hardware architecture used for the execution of the functionalities of this application. These two packages share the same component definition introduced in the `component` package. The `component` and the `factorization` packages are introduced to gather the common concepts used in the different parts of the Y model. The `association` package introduces some basic directives for mapping an application on a given hardware architecture.

The general philosophy of this hierarchy is to define a maximum of common parts for the various aspects of the Y model. The Gaspard2 UML profile is based on a *component oriented* approach which allows to support as much as possible the reuse of software and hardware components. To do that, the different software and hardware elements of the studied system are represented by a *component* structure independently of their environment. We propose here several enhancements allowing to improve and fortify the Gaspard2 UML profile description by introducing some new concepts and OCL constraints [3]. In the following, we present in more details the different packages of the profile (by using the *MagicDraw* UML modeling tool).

`Component` **Package.** The `component` package gathers the common concepts for the application and the hardware architecture metamodels. The main objective of this package consists in defining a support for the component

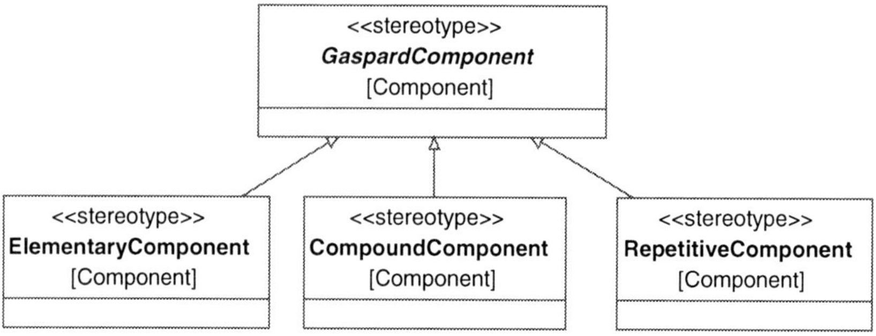

Figure 18.2. `component` package.

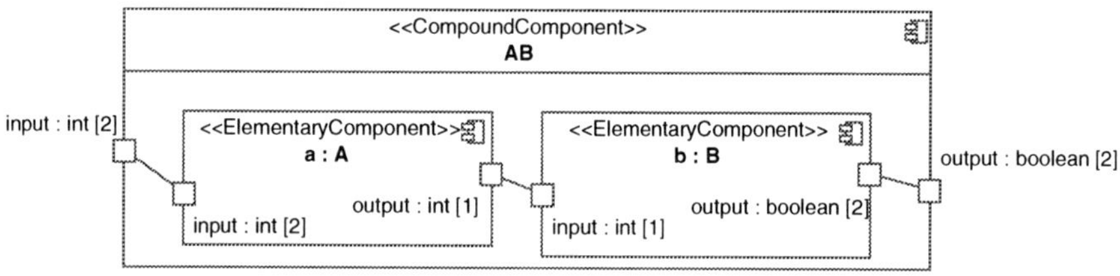

Figure 18.3. Example of a `CompoundComponent` component.

oriented methodology by allowing the definition of a component independently of its environment. Figure 18.2 shows the different concepts defined in the `component` package. In this package, we have introduced three basic stereotypes to specify the Gaspard2 components: `ElementaryComponent`, `CompoundComponent`, and `RepetitiveComponent`. These stereotypes extend the metaclass `component` to specify special elements relating to the Gaspard2 models.

Each Gaspard2 component can be either elementary, compound, or repetitive as expressed by the OCL constraint:

> **inv** : self.stereotype $\rightarrow$ exists (name = ElementaryComponent **xor** name = CompoundComponent **xor** name = RepetitiveComponent)

The `ElementaryComponent` concept represents a particular component which does not have any Gaspard2 component description as specified by the OCL constraint:

> **inv** : self.part $\rightarrow$ select (stereotype $\rightarrow$ exists (name = GaspardComponent))
> $\rightarrow$ size() = 0

In the example of Fig. 18.3, the component A specifies an elementary component which defines a function taking as input a pattern of two integer elements and produce as a result a pattern of only one integer element.

The `CompoundComponent` concept is used to define compound components. This concept can facilitate the hierarchical composition of the various Gaspard2 models, and contains only `GaspardComponent` components as expressed by the OCL constraint:

inv : self.part → exists (p | p.stereotype→exists (name=GaspardComponent))

For example, if the result of the A elementary task is used by another B elementary task which produces a pattern of two boolean elements, then the AB `CompoundComponent` concept is used to group and represent the relation between the two tasks A and B as shown by Fig. 18.3. Figure 18.3 gives a simple example of a `CompoundComponent` contaning two `ElementaryComp-onent` components.

The `RepetitiveComponent` concept allows to describe the repetition of the component modeled inside it. This repetition is defined according to the repetitive concept of the Array-OL model, and represents a repetition field on only one `GaspardComponent` component as specified by the OCL constraint:

inv : self.part → select (stereotype → exists (name = GaspardComponent))
→ size() = 1

Figure 18.4 gives a simple example on the repetition of an elementary task A. In this example, the model receives as input an integer array of two dimensions 6×2, and produces as result an integer array of two dimensions 3×2. Inside the $RepA$ component, the A component is repeated 3×2 times. More details on these concepts are presented in the `factorization` package.

`Application` **Package.** The `application` package, presented in Fig. 18.5, introduces a new stereotype, `ApplicationComponent`, which extends the metaclass `component`, and generalizes the `GaspardComponent` abstract concept.

The `ApplicationComponent` concept refines the `GaspardComponent` concept by adding an applicative connotation. It can be seen as a set of functions which perform calculations on the input data coming from their external environment through input ports, and produce results to their environment through output ports.

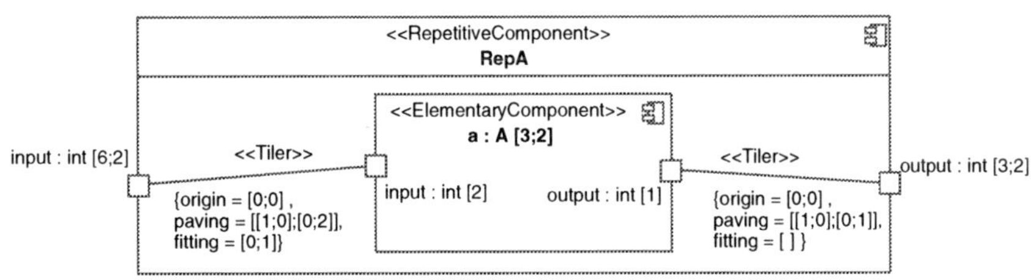

Figure 18.4. Example of a `RepetitiveComponent` component.

`HardwareArchitecture` **Package.** In a similar way to the `application` package, the `hardwareArchitecture` package refines the `GaspardComponent` concept by adding a hardware architecture connotation. This package introduces a new stereotype which extends the metaclass `component` to allow the description of various hardware architecture concepts used as support for the execution of the application. A global view on this package is given by Fig. 18.6.

`Association` **Package.** The `association` package, presented by Fig. 18.7, is used to specify the different allocation mechanisms used to associate an application component to a hardware architecture component. In this package, we can mainly find two kinds of allocation: `DataAllocation` and `TaskAllocation`.

The `DataAllocation` concept is used to map the different data available on the ports onto hardware components such as memory, while the `TaskAlloca-`

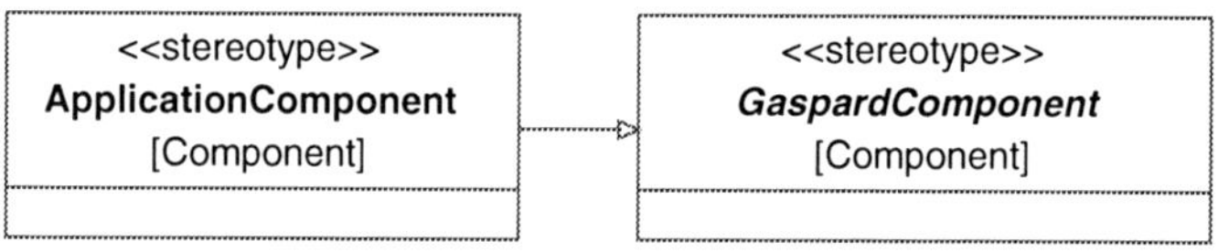

Figure 18.5. `application` package.

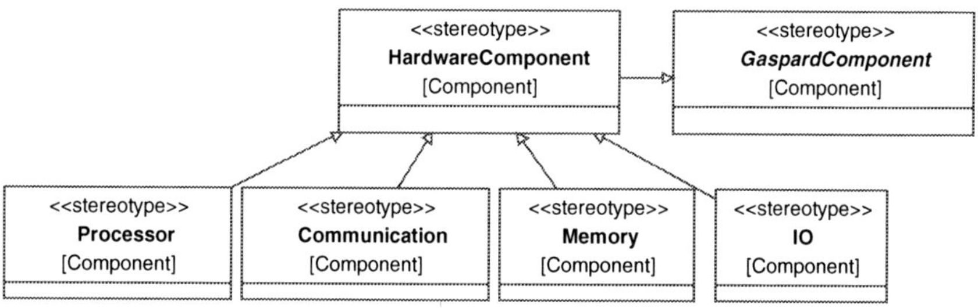

Figure 18.6. `hardwareArchitecture` package.

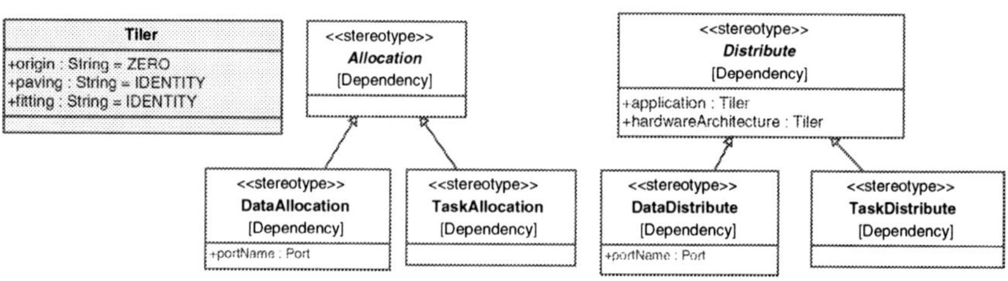

Figure 18.7. `association` package.

`tion` concept is used to map the application tasks onto hardware components such as processor.

The `Distribute` concept is introduced to specify the mapping of a repetitive application onto a repetitive hardware architecture by using Array-OL concepts. This specification is expressed by the following OCL constraint:

inv : self.participant → size() = 2 **and** self.participant → forAll (stereotype → exists (name = RepetitiveComponent))

`Factorization` **Package.** The `factorization` package contains structural factorization mechanisms inspired by the Array-OL model. These mechanisms represent the *multidimensional multiplicity* and the different *link topologies* used to specify the repetitive concept in the Gaspard2 models.

The **multidimensional multiplicity** concept allows to give some directives on the number of element repetitions. The *multiplicity* concept already exists in the UML metamodel, and can be seen as an array of elements with only one dimension. In the Gaspard2 metamodel, we generalize this concept to consider the potentially multidimensional multiplicity. In this case, a Gaspard2 element having a multidimensional multiplicity can be seen as an array of elements with several dimensions[2]. This multidimensional multiplicity concept requires an extension of the UML metamodel and does not appear in the Gaspard2 UML profile description. In the example of Fig. 18.4, the multidimensional multiplicity $[3, 2]$ defined on the component part $a : A$ specifies that this elementary task is repeated 3×2 times. The multidimensional multiplicity on the ports allows to specify multidimensional arrays like in the case of the input port, presented in the same example, which specify an integer array of two dimensions 6×2.

The **Link topology** concept is used to specify an information set associated to the relation between Gaspard2 components. This concept takes the multidimensional multiplicity into account. It allows to express the relation between the different element repetitions, and the Array-OL basic concepts. Figure 18.8 shows the different link topologies defined in the `factorization` package.

The `InterRepetitionLinkTopology` concept is used to specify a special regular link between the different repetitions of the same element as expressed by the OCL constraint:

inv : self.participant → size() = 2 **and** self.participant[0] = self.participant[1]

This kind of topology can be used for example to specify a toric grid topology when each instance of the element is connected to its four neighbors, or in the case of the calculation of a particular function such as:

$$\begin{cases} Y_n \;=\; f(Y_{n-1}, X) \\ Y_0 = initValue \end{cases}$$

In this case, each repetition n of the elementary function f depends on the result provided by the repetition $n - 1$. A simple example of this function is a discreet integral given by Fig. 18.9. The `repetitionSpaceDependence` attribute is used to specify the neighbor position of the element on which the inter-repetition dependency is defined. Since the `InterRepetitionLink-Topology` connection specifies a relation between the different repetition of an element and a uniform dependency, this link must be only defined inside a repetitive component as expressed by the OCL constraint:

inv : self.owner.stereotype → exists (name = RepetitiveComponent)

As shown by Fig. 18.9, the `InputDefaultLink` connector is used to specify the default values of the input port in the case of an inter-repetition dependency. This concept represents a special connector which has only significance when the iteration domain does not give values. In a similar way, the `OutputDefault-Link` connector can be used to assign values to output ports when the iteration domain does not give values.

The `InterRepetitionLinkTopology` and the `DefaultLink` connectors must be defined inside a repetitive component, and between two ports of the same type and multiplicity as specified by the OCL constraints:

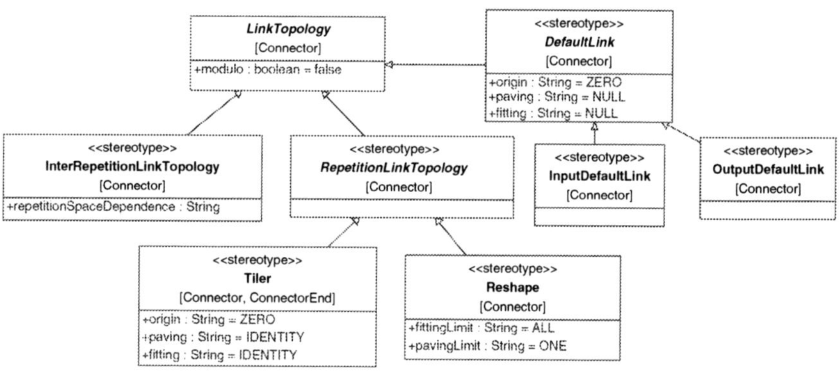

Figure 18.8. `factorization` package.

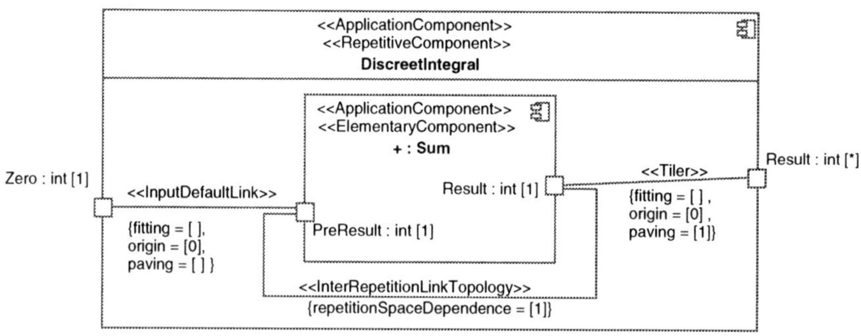

Figure 18.9. Simple example on the use of link topologies.

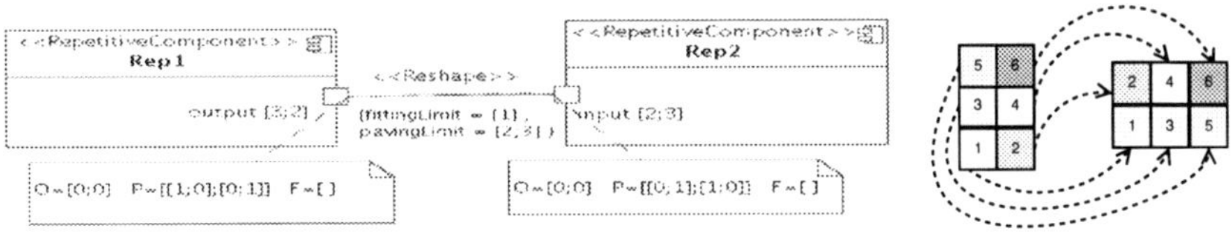

Figure 18.10. Simple example on the use of `Reshape` link.

inv : self.owner.stereotype → exists (name = RepetitiveComponent)

inv : self.connectorEnd → forAll (x,y | x.multiplicity = y.multiplicity **and** x.type = y.type)

Another link topology defined in the `factorization` package is the `RepetitionLinkTopology`. This topology is mainly based on the repetition concepts used in Array-OL models, and mainly defines two connection kinds: `Tiler` and `Reshape`.

The `Tiler` link is used to define a repetitive link between two components according to the Array-OL model. The `origin`, `paving`, and `fitting` attributes are used to specify necessary Array-OL information for the processing of input and output patterns as shown in the example of Fig. 18.9.

The `Reshape` link is used in special cases to specify the redistribution or the reorganization of some array elements on another array as illustrated in the example of Fig. 18.10. To do that, we need to specify Array-OL tiler information on the two ConnectorEnds of the link. The `Tiler` concept must then be defined on the two ConnectorEnds of the `Reshape` connector as specified by the OCL constraint:

inv : self.connectorEnd → forAll (stereotype → exists (name = Tiler))

Moreover, the `Reshape` link has only significance between two repetitive components as indicated by the OCL constraint:

inv : self.participant → size() = 2 **and** self.participant → forAll (stereotype → exists (name = RepetitiveComponent))

Two attributes are defined on the `Reshape` connector: `pavingLimit` and `fittingLimit` which represent respectively the pattern number of used arrays, and the element number of each pattern.

3. Introducing Control in the Data Parallel Application Modeling

As shown above, the construction of the Gaspard2 UML profile is mainly made for systematic data parallel algorithms, and parallel architectures. It allows to describe the data dependencies and all the potential parallelism present in the studied applications. However, this profile does not contain any representation of control or reconfiguration concepts which make difficult, even impossible, the modeling of applications containing control concepts and changing modes. For this reason, we propose to introduce the control concepts in the Gaspard2 UML profile in order to take more general parallel applications, mixing control and data processing, into account. The introduction of control into a parallel application can be done by giving a *reactive behavior* to these applications which has been largely studied in the case of the synchronous reactive systems.

Synchronous Approach for Reactive Systems

Reactive Systems are computer systems that react continuously to their environment, by producing results at each invocation. In the beginning of the 80s, the family of synchronous languages and formalisms has been a very important contribution to the reactive system area [6]. Synchronous languages have been introduced to make programming reactive systems easier. They are based on the *synchrony hypothesis* that does not take reaction time in consideration since each activity is seen as instantanous and atomic.

Synchronous languages, like Lustre, Esterel, and Signal [6], are devoted to the design, programming, and validation of reactive systems. They have a formal semantics and can be efficiently compiled into C code, for instance. Moreover, these formalisms make it possible to validate and verify formally the behavior of the system. In this field, we often speak about tools and approaches for simulation, verification, and code generation for reactive systems specified in a synchronous language. These languages can be *declarative* or *imperative* depending if the system's behavior is respectively mainly regular or discrete. However, the most used embedded systems are generally *hybrid*, and combine control and data processing. For this reason, we need efficient tools and methods taking into consideration this kind of behavior. Declarative or data flow languages, like Lustre, are used when the behavior of the system to be described has some regularity like in signal-processing. Imperative or control flow languages, like Esterel, are more appropriate for programming systems with discrete changes and whose control is dominant. However, the most realistic and used embedded systems have rarely an exclusively regular or discrete behavior. These systems are generally *hybrid*, and combine control and data processing. For this reason, we need efficient tools and methods taking in consideration this kind of behavior.

Related Work

Several approaches have been proposed to facilitate the study of hybrid systems. We can find the *multilanguages* approach which combines imperative and declarative languages, like using Lustre and Argos [7]. However, when using several languages it is very difficult to ensure that the set of corresponding generated codes will satisfy the global specification.

Another design methodology consists in using a *transformational* approach which allows the use of both types of languages for specification but, before code generation, the imperative specifications must be translated into declarative specifications, or vice versa, allowing to generate a unique code instead of multiple ones. This approach is efficient for describing reactive systems combining control and data processing. However, there are systems whose behavior is mainly regular but can switch instantaneously from a behavior to another. The most adapted method to describe this kind of system consists in using a *multistyles* approach which makes it possible to describe with only one language the various *running modes* of the system. The Mode-Automata [8] represent a significant contribution in this field. They are used to clearly express the different running modes of an application and the conditions of switching between modes.

Presented works study the control and data processing combination. However, all these studies do not take the data parallel processing into account. The parallel processing represents generally a set of tasks which can be executed in parallel to define the global behavior of the system like in signal and image processing. In the literature, few works have been proposed to introduce the control in the parallel computation field. For instance, using Signal relational language and the Alpha functional language for the application codesign [9] do not define any specification model allowing the modeling of parallel applications with control concepts. Another example can be found in Ptolemy [10] which proposes a multidimensional computation model (MDSDF) and an automata model (FSM). However, the combination of these two concepts has never been studied.

In our work, we choose to use the Mode-Automata [8] concept and the control/data flow separation methodology [11] to introduce the control and the reactive behavior in the Gaspard2 UML profile. In this study, we are only interested in the control introduction for the Gaspard2 `application` part. This approach allows to have a more readable model since it is based on a modular specification of the different parts of the system. This facilitates the reuse of existing applications, the modification, the introduction, and the deletion of modes. It can also facilitate the simulation and the modular verification of the system.

Introducing Control in the Gaspard2 Application UML Profile

The introduction of control in the Gaspard2 application part is mainly based on the synchronous approach and the control/data flow separation methodology. The introduction of the control into data parallelism applications requires the definition of a *degree of granularity* for these applications [12]. This concept allows to delimit the different execution cycles or *clock signals* in which it becomes possible to take the control values into account. To do that, we introduce new concepts to specify control in Gaspard2 application models. These concepts are gathered in the `control` package.

The `control` package represents the basic elements used to express an automaton structure and the different running modes of the studied system. The question which arises now is: *how to model the control automaton and the different running modes in the Gaspard2 UML profile?*

Modeling the control automaton. The control automaton structure that we propose to use is inspired by the one of Mode-Automata [8] and Moore-Automata. The use of the Mode-Automata concept allows to well specify the different running modes of the system and the switch conditions between modes, while the Moore-Automata concept is used to limit the output of the control automaton to its current state. In this case, the transition function only specifies a set of events that allow the switch between states without having any effect on the output result of the automaton.

To introduce control concepts in the Gaspard2 application models, we must take the Gaspard2 model semantics into account. The Gaspard2 semantics do not express any flow concept since the time is represented by an unordered and infinite dimension. However, in the control automaton structure, the flow concept is present and significant for the definition of its behavior. For this reason, it becomes important to propose a model representation of the automaton structure by respecting as much as possible the Gaspard2 model semantics.

A possible solution consists in using a dependency relation between the different repetitions of the automaton transition function. This dependency makes it possible to memorize the previous state of the automaton and then to respect the general automaton structure semantics. To model this dependency relation, we use the `InterRepetitionLinkTopology` concept already existing in the Gaspard2 UML profile definition. This link topology allows the introduction of the flow concept in the Gaspard2 models, and makes possible the modeling of an automaton structure without changing or modifying the basic Gaspard2 semantics. To introduce the automaton structure in the Gaspard2 UML profile, we define new concepts in the `control` package as shown by Fig. 18.11.

The `TransitionComponent` abstract stereotype is used to specify the transition function of an automaton, while the `RepetitiveTransitionComponent` stereotype is used to define the repetition of the transition function. To model the repetition on the transition function, an `InterRepetitionLinkTopology` connection must be specified on the different repetitions of the transition function as specified by the OCL constraint:

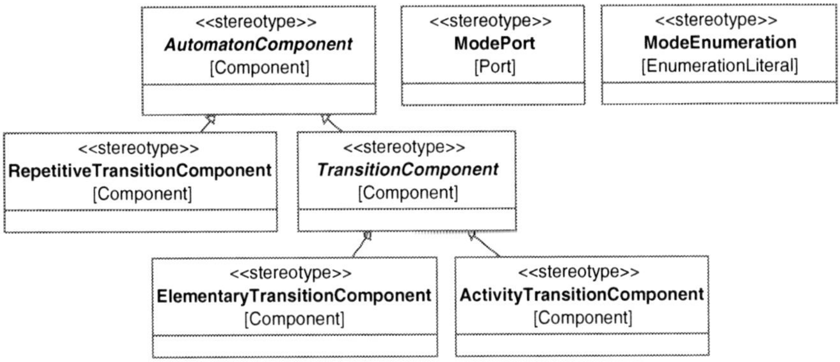

Figure 18.11. Gaspard2 UML modeling for the control automaton.

inv : self.connector $\rightarrow$ select (stereotype $\rightarrow$ exists (name = InterRepetitionLinkTopology) $\rightarrow$ size() = 1)

The description of the automaton transition function can be done in two different ways: by using an `ElementaryTransitionComponent` or an `ActivityTransitionComponent` components. In the case of an `ElementaryTransitionComponent`, the behavior of the automaton transition function is regarded as a black box without using any Gaspard2 component description as specified by the OCL constraint:

inv : self.part $\rightarrow$ select (stereotype $\rightarrow$ exists (name = GaspardComponent)) $\rightarrow$ size() = 0

In the case of an `ActivityTransitionComponent`, the behavior of the transition function must be specified by a UML activity diagram as specified by the OCL constraint:

inv : self.feature $\rightarrow$ forAll (oclIsKindof (Activity)) **and** self.part $\rightarrow$ size() = 0

Other OCL constraints are also added to the profile definition to express the link between the transition function and the repetition on this function. For example, each `RepetitiveTransitionComponent` component must contain only one `TransitionComponent` component as expressed by the OCL constraint:

inv : self.part $\rightarrow$ size() = 1 **and** self.part.stereotype $\rightarrow$ exists (s | s.name = ElementaryTransitionComponent **or** s.name = ActivityTransitionComponent)

and each `TransitionComponent` component must be defined inside a `RepetitiveTransitionComponent` component as specified by the OCL constraint:

inv : self.owner.stereotype $\rightarrow$ exists (name = RepetiveTransitionComponent) **and** self.owner $\rightarrow$ size() $\geq$ 1

To specify the execution mode, the `AutomatonComponent` abstract concept must have only one output `ModePort` port of `ModeEnumeration` type as expressed by the OCL constraint:

inv : self.ownedPort.required → size() = 1 **and** self.ownedPort.required.stereotype → exists (name = ModePort)

This port provides the *mode value* used by the calculation part to choose the running mode to activate among several exclusives.

Modeling the different running modes. The output mode value provided by the control automaton is used by the calculation part controlled by this automaton. This calculation part represents the various running modes of the studied system, and allows the activation of the corresponding mode.

The controlled part represents a special component having a *switch* behavior. To model this part in the Gaspard2 application profile, we introduce a new Gaspard component that we called `ControlledComponent` (Fig. 18.12). This component specifies the switch between several exclusive calculation parts, and must have one input `ModePort` port used for the activation of the corresponding calculation mode as expressed by the OCL constraint:

inv : self.ownedPort.provided → select (stereotype → exists (name = ModePort) → size() = 1)

To model the different running modes, we introduce a new concept `AlternativeComponentPart` which is only applied to the UML component `part`. This concept gives a particular behavior to the component parts by specifying that they can be activated only if their `activationCondition` value is equal to the mode value. In other words, the controlled component will only contain the alternative component part to activate as expressed by the OCL constraint:

inv : self → includes (p:part | p.activationCondition.value = ownedPort.provided → select (stereotype → exists (name = ModePort)).value) **and** self → excludes (p:part | p.activationCondition.value <> ownedPort.provided → select (stereotype → exists (name = ModePort)).value)

To express the switch behavior, the controlled component contains only alternative component parts as expressed by the OCL constraint:

inv : self.part → forAll (stereotype → exists (name = AlternativeComponentPart))

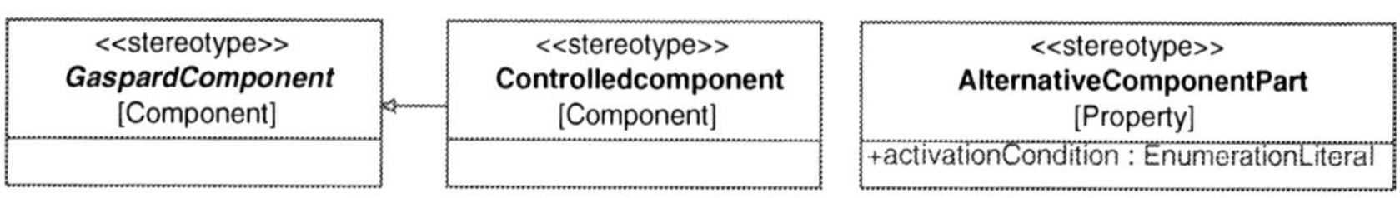

Figure 18.12. Gaspard2 UML modeling for the controlled part.

The different alternative parts of a controlled component must have the same interface to facilitate the reuse and the manipulation of the existing components as expressed by the OCL constraint:

> **inv** : self.part → forAll (p1,p2 | (p1.ownedPort.provided = p2.ownedPort.provided) **and** (p1.ownedPort.required = p2.ownedPort.required))

To certify the correct functioning of the controlled part, the various alternatives components must be exclusive as specified by the OCL constraint:

> **inv** : self.part → select (taggedValue.name = activationCondition **implies** forAll (p1, p2 | p1.taggedValue.value <> p2.taggedValue.value))

Figure 18.13 gives a simple example on the modeling of a controlled application. In this example, the system can operate in two different modes $AMode$ and $BMode$. The activation and the switch between these two modes are done by the control automaton according to the `event` value. This example shows the relation between the control part and the controlled calculation part through the mode value. It also specifies the modeling of the different alternative modes having the same interface.

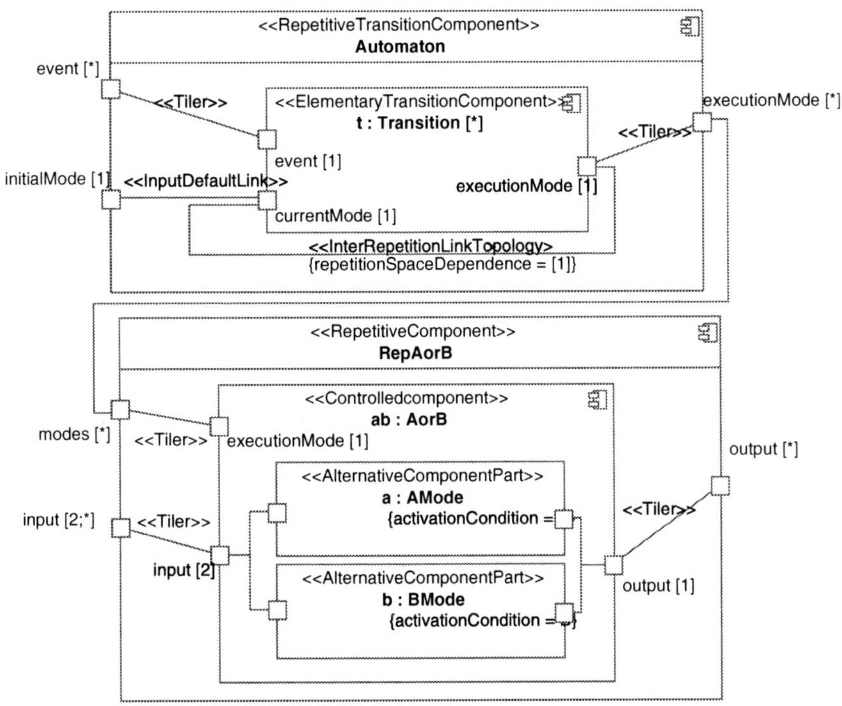

Figure 18.13. Simple example on the modeling of a controlled application.

4. Conclusion and Future Work

In this paper, we have proposed a UML2 profile allowing the introduction of the control concepts in the Gaspard2 application profile. The proposed profile is based on a clear separation between control and data parallel parts. It respects the concurrency, the parallelism, the determinism, and the compositionality of the studied systems.

The main goal of our work consists in proposing a UML solution for modeling the control automata, the different running modes of an application and the link between the control and the computation parts in the case of a parallel applications. This specification allows to study more general parallel systems mixing control and data parallel processing. Our profile proposition, notably hardware architecture and association parts, participates in the definition of ProMARTE profile [4]. In future work, we will propose the introduction of the control concepts in the architecture and the association Gaspard2 profiles allowing to take the configurability concept into account for the architecture models, and to have a better use of the mapping and scheduling algorithms for the association models. There is also ongoing work to introduce clock concepts in Gaspard2 models. This study will allow us to use UML state machine to model the control part.

Acknowledgment

The authors would like to thank WEST team member for their contributions in the definition of the profile presented in Section 2.0.

Notes

1. http://www.lifl.fr/west/gaspard
2. The Gaspard2 elements which can have a multidimensional multiplicity are only *ports* and *parts*.

References

[1] Philippe Dumont and Pierre Boulet, *Another Multidimensional Synchronous Dataflow: Simulating Array-OL in Ptolemy II*, Research Report RR-5516, INRIA, March, 2005, www.inria.fr/rrrt/rr-5516.html

[2] Arnaud Cuccuru, Jean-Luc Dekeyser, Philippe Marquet, and Pierre Boulet, *Towards UML 2 Extensions for Compact Modeling of Regular Complex Topologies - A Partial Answer to the MARTE RFP*. In: MoDELS/UML 2005, ACM/IEEE 8th International Conference on Model Driven Engineering Languages and Systems, Lecture Notes in Computer Science **3713**, pp. 445–459, Montego Bay, Jamaica, October, 2005

[3] Object Management Group, *OCL 2.0 Specification, Version 2.0*, June, 2005, http://www.omg.org/docs/ptc/05-06-06.pdf

[4] ProMarte partners, *A UML Profile for MARTE, Initial Submission*, version 0.44, November, 2005, realtime/05-11-01, www.omg.org/cgi-bin/doc?realtime/2005-11-01

[5] Arnaud Cuccuru, *Modélisation unifiée des aspects répétitifs dans la conception conjointe logicielle/matérielle des systèmes sur puce à haute performances*, Ph.D thesis, Laboratoire d'Informatique Fondamentale de Lille, Université des Sciences et Technologies de Lille, September, 2005

[6] Nicolas Halbwachs, *Synchronous Programming of Reactive Systems.* Kluwer Academic Publishers, Norwell, MA. ISBN 0-7923-9311-2, 1993

[7] M. Jourdan, F. Lagnier, F. Maraninchi, and P. Raymond, A Multiparadigm Language for Reactive Systems. *IEEE International Conference on Computer Languages (ICCL)*, Toulouse, France, 1994

[8] *Florence Maraninchi and Yann Rémond*, Mode-automata: about modes and states for reactive systems, *European Symposium On Programming*, Springer, LNCS 1381, Lisbon, Portugal, March, 1998

[9] Irina Madalina Smarandache, *Conception conjointe des applications régulières et irrégulières en utilisant les langages Signal et Alpha*, Ph.D Thesis, Université de Rennes 1, October, 1998

[10] Edward A. Lee, *Overview of the Ptolemy Project.* Technical Memorandum UCB/ERL M01/11, University of California, Berkeley, March, 2001, http://ptolemy.eecs.berkeley.edu

[11] Ouassila Labbani, Jean-Luc Dekeyser, and Pierre Boulet, Mode-automata based methodology for Scade, In: *Springer, Hybrid Systems: Computation and Control*, 8th International Workshop, LNCS series, pp. 386–401, Zurich, Switzerland, March 2005

[12] Ouassila Labbani, Jean-Luc Dekeyser, Pierre Boulet, and Éric Rutten, *Introducing Control in the Gaspard2 Data-Parallel Metamodel: Synchronous Approach*, International Workshop MARTES: Modeling and Analysis of Real-Time and Embedded Systems (in conjunction with 8th International Conference on Model Driven Engineering Languages and Systems, MoDELS/UML 2005), Montego Bay, Jamaica, October 2005, http://www.lifl.fr/west/publi/LDBR05.pdf

Chapter 19

MCF

A Metamodeling-based Visual Component Composition Framework

Deepak A. Mathaikutty and Sandeep K. Shukla
FERMAT Lab, Virginia Tech, Blacksburg, USA
mathaikutty@vt.edu; shukla@vt.edu

Abstract Reusing IP-cores to construct system models facilitated by automatic generation of the glue-logic, and automated composability checks can help designers to create efficient system level models quickly and correctly for faster design space exploration. With the rise of multiple transaction level (TL), and register-transfer level (RTL) abstractions, constructing models with mixed abstraction levels is also important for designers. This chapter presents a framework that allows designers to (i) describe the structure of components, their interfaces, and their interactions with a semantically rich visual front end, (ii) automated selection of IPs from a component library based on sound type theoretic principles, and (iii) constraint based checks for composability is highly desirable in this context. Furthermore, using an XML based schema to store and process meta-information about the IP models, as well as the schematic visual model helps the process of IP selection and composition. With these in mind, we present MCF, a metamodeling-based component composition framework for SystemC based IP core composition at multiple and mixed abstraction levels, with all the advantages stated above.

Keywords Metamodel, metamodeling framework, architectural template, component composition framework, IP composition, reuse

1. Introduction

One recent industrial trend in system level design is to use high-level programming languages such as C++ to build custom design exploration and analysis models, especially with the increased usage of SystemC. The proliferation of SystemC IP cores necessitates CAD frameworks that support IP core composition to build system models. One of our goals is to reduce much of the

S.A. Huss (ed.), Advances in Design and Specification Languages for Embedded Systems – Selected Contributions from FDL'06, 319–337.

software engineering (SE) and C++ programming burdens that a designer has to face in creating models from existing cores. Some of the SE problems arise due to strong typing requirements, lack of a rigorous compositional semantics of SystemC, and the ad hoc mode of reusability associated with the compile-link-test methodology. Therefore, we need a framework that allows designers to architect the system in terms of components and their composition. Such a component composition framework (**CCF**) makes the designer oblivious to the specifics of the software language and allows realizing a system through *design templates* and *tool support*. The design template may structurally reflect the modeling abstractions such as register-transfer level (**RTL**), transaction level (**TL**), mixed-abstractions (**RTL-TL**) and structural hierarchy. The tools would facilitate selection and connection of the correct components, automatic creation of correct interfaces and transactors, simulation of the composed design and finally testing and validation for correctness of composition.

In Figs. 19.1 and 19.2, we illustrate the use of *design templates* through examples such as the RTL model of the AMBA AHB and TL model of the simple bus from the SystemC distribution. The AMBA bus model shown in Figure 19.1 consists of three masters and four slaves with similar interface descriptions that communicate through multiplexed signals selected by an arbiter and a decoder. For creating this model as a description and later instantiating as an executable using reusable components, the CCF should allow for the following in its description process: (i) generic RTL components that would be reused for the masters and slaves; (ii) generic mux and demux with automatic interface inference for (de-)multiplexing; (iii) RTL components that describe the arbiter and decoder; and (iv) signal-level connectivity of the components.

Figure 19.2 describes the TL model, which consists of three bus masters with distinct protocol-level interfaces (blocking, non-blocking, and direct read-write

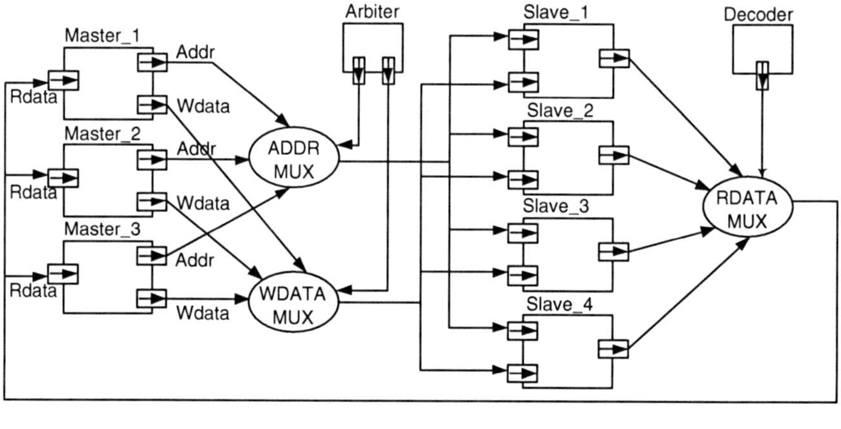

Figure 19.1. RTL AMBA AHB bus model.

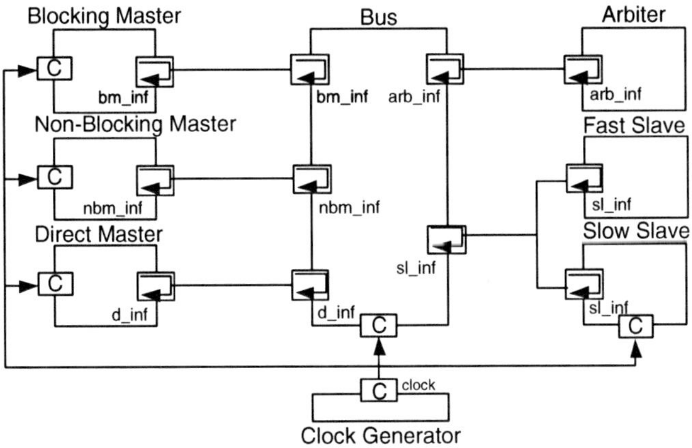

Figure 19.2. TL simple bus model.

access) and an arbiter to schedule their transactions. It consists of two slaves, one with instantaneous read-write access and the other with slow read-write access due to its clock-based activation. For this the framework should facilitate the following in its description process: (i) TL components with function calls that describe the distinct masters, (ii) a bus channel that implements the read-write interface of the masters, (iii) an arbiter channel that implements the arbitration interface, (iv) a channel that implements the read-write interface for the fast slave and which is reused to describe the slow slave, (v) insertion of clock ports for the components and channels, and (vi) transaction-level connectivity.

If the designer wants to explore the simple bus model to analyze various trade-offs, then an alternative description could be replacing the slave channels with RTL slaves from the AMBA bus in order to determine the loss in simulation efficiency due to the gain in precise timing from the extra handshaking introduced. For such an alternative the framework's description process should allow: (i) modification through reusability of slave components from the AMBA bus and (ii) insertion of transactors that map the TL and RTL interaction of the bus channel and the RTL slaves. Furthermore for the three scenarios explained above, the framework in general should allow design visualization and enforce checks that look for inconsistencies, datatype mismatches, incompatible interfaces, and illegal component or channel description.

Once such checks are done, the CCF *tool support* should provide automatic selection of IPs at different abstraction levels, code generation for insertion of transactors to bridge the abstraction gap, testbench generation, and composing all the above to arrive at a simulation model. The selection is performed at both the structural as well as the behavioral level to obtain all plausible implementations that match the CC model.

We define the problem statement in terms of what is needed for a CCF: (i) visual language for describing component composition models; (ii) modeling abstractions such as RTL, TL, and mixed; (iii) formal constraints to enforce different kinds of checks to catch discrepancy in the model; (iv) consistency check to ensure system-level design constraints; (v) type checking and propagation to automate type inference; (vi) intermediate representation of the CC model; (vii) library of compiled C++ objects of different IPs; (viii) reflection of structural and behavioral types of these IPs, and (ix) introspective composer that performs IP selection and connection, insertion of bridges, validation of functionality, and produces an executable model.

Design Flow

An abstract view of the design flow is shown in Fig. 19.3, where the user begins by creating a component composition model using the CC language. Upon conformance to the various constraints and completion of the model, it is converted to an intermediate representation (IR), which is given as input to our introspective composer (IComp). On the other hand, we have a library of compiled objects of C++ IPs from which the meta-level information (structure, timing, etc.) is extracted through our automated reflective mechanism and expressed into XML. This XML library is also given as input to the IComp, which is responsible for constructing an executable model by querying the reflected database and selecting the appropriate IPs, inserting the transactors and composing them. The main focus of this paper is the CC language and model.

Main Contributions

Our main contributions in this paper can be summarized as follows: (i) a visual language (**CCL**) and framework for component composition using

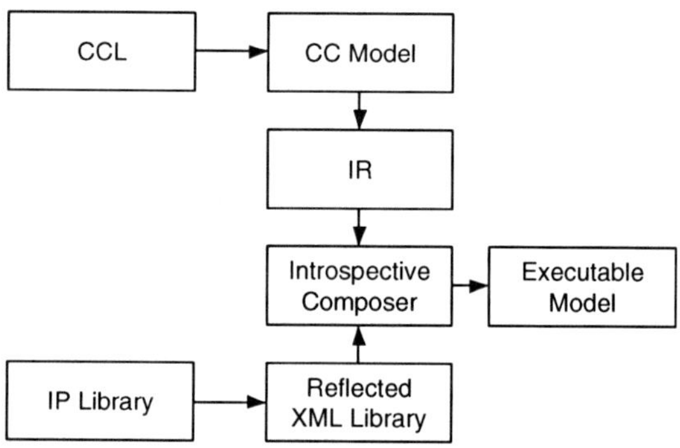

Figure 19.3. Abstract design flow.

the metamodeling paradigm of the generic modeling environment (GME) [5]; (ii) design-time constraints on objects and connections in a CC model based on the hardware design language (HDL) and software requirements expressed using the Object Constraint Language (OCL); (iii) consistency checker that performs static analysis of the CC model; (iv) type checker that performs datatype checking; (v) type propagation engine that derives the interface types based on different type inference algorithms used; and (vi) translation of the CC model into an intermediate XML representation that will be given as input to the introspective composer. The metamodeling framework of GME allows representing the syntax, semantics, and constraints into visual language and modeling framework. Various checkers and engines are developed using GME APIs and integrated into the modeling framework as analyzers.

Organization

The rest of the paper is organized as follows: In Section 2, we briefly outline the related work and the following section introduces the background and some of the relevant terminologies. Section 4 describes the CC model description with examples. In Section 5, we briefly describe the consistency checker, type checker, type propagation engine, and the XML translation. Finally, we conclude by summarizing our modeling experience and discussing our future work.

2. Related Work

In this section, we briefly introduce related work on CCF and distinguish ourselves. Balboa [3, 2] is a framework to compose components and is not based on a metamodeling framework. Components are connected and their interface specified via the component integration language (CIL). Before beginning simulation, the framework performs type inference to select appropriate implementation based on the port-connectivity. The CIL is a script-like language used to manipulate and assemble C++ components with constructs that provide an abstraction over C++. Our CCL is built on top of a UML-based visual language therefore, it is rich in object oriented features and has formalism attached that enforces HDL constraints on the model and allows for structural hierarchy. Our checkers perform different kinds of static analysis and inference to remove inconsistencies. As a result, our CCL is far more expressive than the CIL. Furthermore, we envision MCF to select components based on behavioral types of the components in the system. Furthermore, our CCL provides design templates for various abstraction levels as opposed to Balboa, which is meant for RTL. Our component reuse methodology is distinct (uses an automated reflective mechanism based on XMLized extraction of meta-info) and much simpler to use than BALBOA's ad hoc methodology (user-written interface descriptions called BIDLs).

SPIRIT [8] enables system design using a generic metadata model [6] that captures configuration and integration-related specifics (*interoperability metadata*) of an IP into a XML library. This enables an SoC design flow in which the modeler instantiates and characterizes these IPs to create the system design. The modeler is required to perform the tedious characterization process, which could be automated by a CCF through selection and composition techniques. As a part of the SoC design flow, SPIRIT provides API standards that should be implemented to export the design and configure the actual IPs to create the executable. Therefore they do not have a tool, instead provide the standards for the metadata capture and the configuration and integration APIs. For interoperability of MCF and SPIRIT-based tools, we plan to develop a generic metamodel that imports the SPIRIT metadata and implements the respective APIs as plugs-ins to facilitate the IP integration, as part of our CCF.

The Liberty Simulation environment [11] is a framework for construction of micro-architectural simulation models. A model is structural and concurrent consisting of a netlist of connected instances of components, which communicate through ports. The Liberty framework has a language similar to our CCL, but is devoid of any formalism and the main distinction being that our metamodel driven formalism is rigorous and extensible. They perform type inference and support polymorphic types for their components [10], which is also handled as a part of our modeling framework. Our modeling language is abstract and does not associate any simulation semantics with the components, but this is part of our future work. Furthermore, they do not allow different modeling abstraction besides RTL.

Ptolemy II is a modeling environment designed to support multiple models of computation. It has a type system [12] that supports type inference in the presence of parametric polymorphism and subtyping. Metropolis [9] is an example of a metamodeling-based design environment for embedded systems. Vendors are beginning to support higher-level synthesis with behavioral modeling and automatic IP selection and insertion [4].

3. Background

Metamodeling A *metamodeling framework* facilitates the description of a modeling domain by capturing the domain-specific syntax and static semantics into an abstract notation called the *metamodel*. This metamodel is used to create a modeling syntax that the designer can use to construct domain-specific models by conforming to the metamodeling rules governed by the syntax and static semantics of the domain. A framework that enables this is called a *modeling framework*. Static semantics refer to the well-formed-ness of syntax in the modeled language, and are specified as invariant conditions that must hold for any model created using the modeling language. We identify two distinct roles played in a metamodeling framework. The role of a meta-user who constructs domain-specific metamodels and the role of a user who instantiates these metamodels for creating domain-specific models.

Generic Modeling Environment (GME) The Generic Modeling Environment (GME) [5] is a configurable toolkit that facilitates the easy creation of domain-specific modeling and program synthesis environment. The metamodeling framework of GME provides a set of generic concepts, which are abstract enough such that they are common to most domains. These concepts can then be customized into a new domain such that they support that domain description directly. The generic concepts describe a system as a graphical, multiaspect, attributed entity-relationship (MAER) diagram. Such an MAER diagram is indifferent to the dynamic semantics of the system, which is determined later during the model interpretation process. The generic concepts supported are *hierarchy*, *multiple aspects*, *sets*, *references*, and *explicit constraints*. The models created are either used to generate applications or to synthesize inputs to different COTS analysis tools.

Preliminary Definitions

A ***leaf*** component is a representation of a hardware block without embedding other components. A ***hier*** component is a hierarchical hardware block that embeds other leaf or hier components. The interface of these components differs based on the level of abstraction. The different modeling abstractions provided are the RTL, where components contain input–output (IO) ports that describe their interface behavior and use signals. In TL abstraction, the components interact through read-write function calls and use channels to dictate the transaction at the interface. Furthermore, we provide a mixed abstraction, where component at RTL and TL communicate through transactors that bridge the two abstractions. The modeler instantiates a component and then characterizes it by specifying the attributes and defining its internals. The allowable characterizations and possible containments of an object and the relations that it can participate in, defines its ***static requirement*** which is fulfilled by the modeler and enforced by the modeling framework. The ***type-contract*** of a component describes the type bindings of its interface and internals. The ***type bindings*** at the RTL are the datatypes specified on the input and output ports and at TL are the datatypes specified on the parameters of the read and write function calls. A component is ***partially typed***, when the type-contract consists of one or more unspecified (UN) type bindings. A component is ***untyped*** when all the type bindings in the type-contract are UN. Two components are said to be ***structurally equivalent*** when their type-contracts can be related through some notion of equivalence.

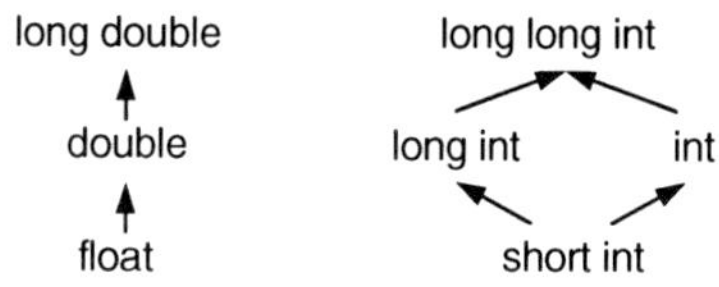

Figure 19.4. Part of the C++ type-structure.

In order to allow for extensive type checking and type propagation, we represent the basic types of the design domain into a ***type-structure*** that is given as input to the different checkers and engines. Figure 19.4 shows a part of the type-structure for C++. The structure is used to perform two types of checking defined below:

Definition 3.1. Type-structure: It is a directed acyclic graph, which represents the hierarchy for the basic types supported by the design language.

Definition 3.2. Type equivalence: Two types are said to be type equivalent, if they map to the same node of the type-structure.

Definition 3.3. Type subsumption: A type t_i is a subsumption of t_j if both belong to the same connected subgraph of the type-structure T and t_i is a child of t_j or both have a common parent.

4. CC Metamodel and Model

The design environment is built on top of a GME [5] and captures the syntax and semantics of a visual composition language (CCL) into an abstract notation called the ***CC metamodel*** **(CCMM)**. It allows different design templates that facilitate a description in terms of component instantiations, characterization, and their composition that we call the ***CC model*** **(CCM)**. The CCMM is developed using UML class diagrams and OCL constraints which is translated into visual syntax by GME [5]. The visual design environment allows a modeler to construct codesign models conforming to the underlying semantics of the CCMM.

CCMM is represented as $CCMM = \langle E, R, E_C, R_C, A\rangle$ where, E is a collection of entities and R is a collection of relationships that relate two objects. E_C is a collection of constraints described on the entities, whereas R_C is a collection of constraints on the relationships in the metamodel. Real-world objects are often represented and analyzed in different ways (e.g. multiple views). Similarly, a metamodel can also have multiple aspects, where each aspect reveals a different subset of entities and their relations. A is the set of aspect that defines the possible visualizations for the CCMM given by:

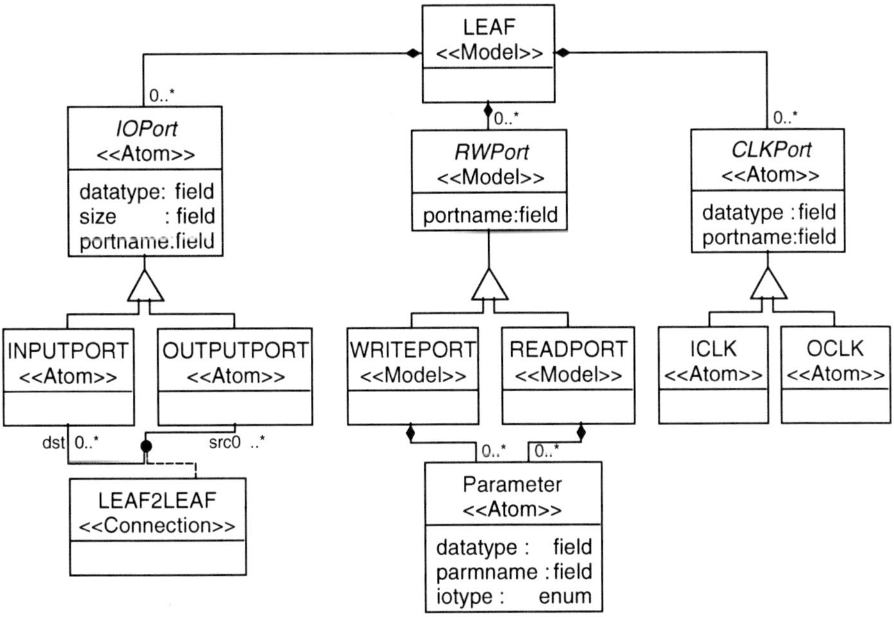

Figure 19.5. UML Diagram of a LEAF Object.

$$\text{A} = \{\text{DesignView, ComponentView, Point2PointView, ChannelView, MixedView, BusView, NetworkView}\}$$

Where, for example *Point2PointView* is meant for RTL-specific visualization and *ChannelView* is for TL-specific visualization.

The three main entities in the CCMM are LEAF, HIER, and TRANSACTOR. We discuss an entity and a relation defined in the CCMM and the complete description of the CCMM grammar is provided in [1]. Figure 19.5 shows the class diagram of a LEAF entity that acts as a containment of three kinds of objects namely IOPort, RWPort and CLKPort. The IOPort is of two types INPUTPORT and OUTPUTPORT that represents the input and output ports of an RTL component. The static nature of such a port is captured through attributes namely *portname* and *datatype*. Certain other requirements are expressed through the entity relationships. The LEAF2LEAF <<Connection>> is one such relation that associates two instances of the LEAF entity, which represents the communication between two components. It should be noted that an IOPort does not have an internal structure.

The second type of object a LEAF can contain is a RWPort, which is of two types READPORT and WRITEPORT that describes the read-write function calls of a TL component. These objects have an internal structure that allows the object to act as a container for objects of type Parameter. These internals capture the static requirements of read-write function calls namely function arguments and return type. The attributes that characterize a Parameter are *parmname*, *datatype*, and *iotype*. The third object that a LEAF can contain is a CLKPort, which is also of two types namely ICLK and OCLK, which denotes the clocking behavior of an RTL/TL component. The *datatype* attribute of a CLKPort is predefined with an unalterable value *boolean*.

During modeling, the user begins by instantiating the LEAF entity to describe a component and proceeds to describe the internals. Based on whether the user wants an RTL or TL behavior, the internals would be described by instantiating one or more IOPort or RWPort objects. These internal are then characterized by filling in the attributes, which describes the type of data communicated through these ports. Furthermore, the component can be clocked by instantiating CLK-Port objects. Even though the LEAF instance in the CCM can contain three different types of objects, at model time it is not allowed to have both IOPort and RWPort objects, since a component can either be at RTL or at TL. This static requirement is enforced through OCL constraints in the CCMM. An example of an abstract RTL CCM specified using the CCL is shown in Fig. 19.6, where the blocks with the centered 'L' are LEAF instances with IOPort objects (black squares) and CLKPort objects (red squares).

Therefore, an entity is the encapsulation of a set of associated objects that is generic enough to be seen as a template for a hardware component. For example, a LEAF entity structural represents any elementary hardware component specified at the RTL or TL abstraction. A HIER entity is used to represent a

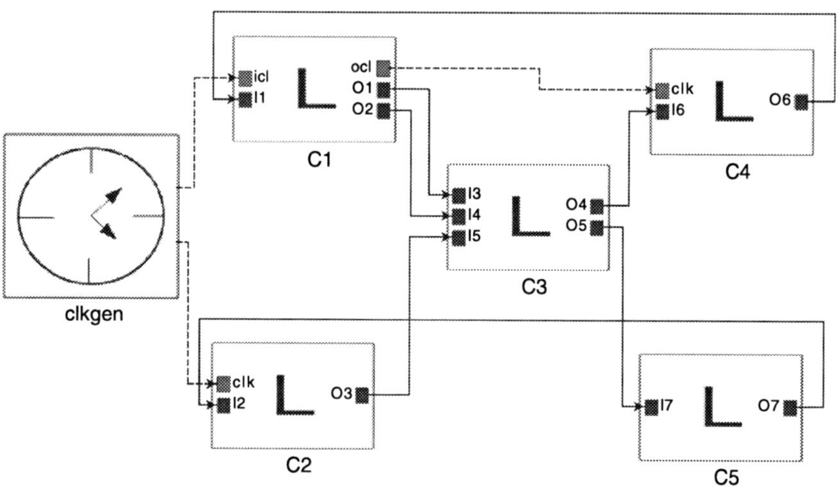

Figure 19.6. An abstract RTL CCM example.

hierarchical hardware component with an internal structure consisting of other LEAF or HIER instances. Finally a TRANSACTOR entity represents a component at mixed RTL-TL abstraction and is used to model components that interface RTL and TL components.

A CCM is the instantiation and connection of entities from the CCMM to create an architectural template of the system. It is represented as $CCM(a)$[1]$= \langle d, I, I_C, C, C_C \rangle$, where d describes the design style, which is one of the following RTL, TL, or RTL-TL (mixed). I is a set of instances of entities from the CCMM. The entry I_C collects the activated constraints on the instances, which are enforced by CCMM as *metamodeling rules* and checked during design-time (instantiation and characterization). A metamodeling violation will undo the incorrect modeling step performed by the user, thereby promoting a design process that is correct-by-construction. C is a set of connections, where a connection associates two objects through a relationship defined in the CCMM. We use connections in our CCM primarily for composition of entities such as LEAF2LEAF <<Connection>>. C_C collects the activated constraints on the connections, which are also enforced by CCMM and checked during design-time while connecting two entities. Furthermore, the parameter 'a' is for rendering a particular visualization for CCM and given a value from A.

The instances in the CCM are grouped into three distinct categories namely components (C_E), medium (M_E), and clock generators. Every instance of an entity is associated with a *name* that is predefined with the name of the entity been instantiated. The user characterizes this instance with a unique name as a part of its description (C1, C2, etc. in Fig. 19.6) unless an already existing instance is being reused. C_E is the collection of LEAF, HIER, and TRANSACTOR instances, where every LEAF instance is a collection of input-output ports or read-write function calls and clock ports. Each input/output port is characterized by their attributes namely *portname* (id), *datatype* (dt), and *size* (sz).

Let T be the type-structure, then $dt \in T$ and $sz \in N$. The *size* attribute is used to allow a port encapsulation. Consider an example of a 2-input adder being described using an instance of the LEAF entity. It has two input ports (i_1 and i_2) and two output ports that provide the sum and carry. Characterization of an integer datatype input port (i_1) of size 1 is shown below:

$$f_{id}(INPUTPORT, i_1),\ f_{dt}(i_1, integer), \text{ and } f_{sz}(i_1, 1)$$

The functions f_{id}, f_{dt}, and f_{sz} are used to assign the attributes with user-defined values such as the function f_{dt} that takes two arguments namely an object and a value and places the value in the *datatype* attribute of the object. At model time, the user selects the instantiated entity that needs to be characterized within the visual framework and textually specifies the values for the attributes of the entity in the fields attached to the instance.

The TL description of a leaf has read-write function calls, where each of these have a set of parameters that represent function arguments, which are read from or written to through these function calls during a transaction. A read interface allows access to its values, whereas a write interface modifies its values. These parameters are characterized by the following *parmname*, *datatype*, and *IOtype* (iotype) attributes. The characterization of a read function call is shown below:

f_{id}(READPORT, read_data)
  f_{id}(read_data.Parameter, data) and f_{id}(read_data.Parameter, addr)
    f_{dt}(data, *string*) and f_{iotype}(data, output)
    f_{dt}(addr, *integer*) and f_{iotype}(addr, input)

This example instantiates a read function call $read\_data$, with two parameter values $data$ and $addr$. The $data$ is a string parameter that is output by the $read\_data$ port given an input for the $addr$ parameter.

M_E is a collection of communication primitives used for different styles of communication. The different communication entities allowed through the CCMM are Splitter, Merger, Channel, Switch, and Bus. The Splitter and Merger media are used for (de-)multiplex RTL-level components, whereas Channel is an abstract channel media for TL communication. An example of an abstract TL CCM communicating through a channel (Channel_0) specified using the CCL is shown in Fig. 19.7, where the blocks are LEAF instances with RWPort objects (green squares). Some of the communication mediums are allowed to have both RTL and TL based description, similar to C_Es. The entities Bus and Switch that provide bus-based and network-on-chip (NoC) style communication are examples of such multi-purpose communication primitives.

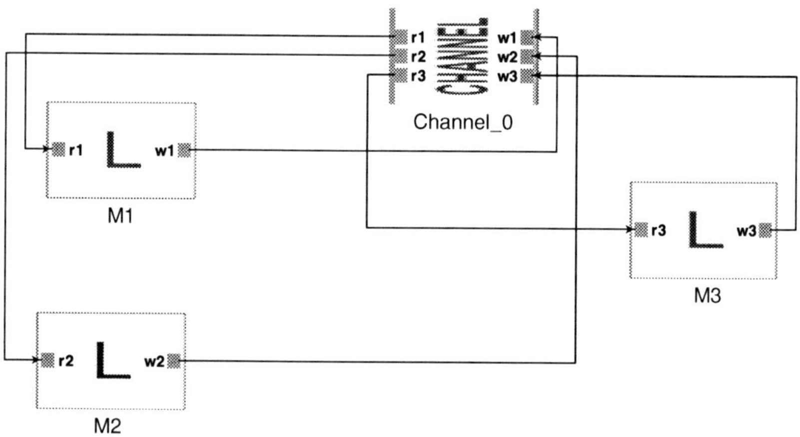

Figure 19.7. An abstract channel-based TL CCM example.

A channel is a mode of communication that implements a set of abstract interfaces to facilitate implicit data and control transfer. Every interface that a channel inherits has a set of virtual read-write functions that is implemented through the channel. Therefore, a channel has a set of concrete read-write function implementations for the methods defined through the different interface inherited by the channel. It is also allowed to have clock ports to describe the clocking nature of the channel. In our metamodel, the inheritance relationship between the interface and the channel is described by a set of <<CONNECTION>>s. A connection between an abstract port p_{intf} of the interface and a concrete port p_{ch} of the channel implies that port p_{intf} is being implemented by the channel through p_{ch}.

In Fig. 19.8.b, we illustrate the block diagram of the FIFO modeled using our CCL. The FIFO (*fifo*) is described as an instance of a CHANNEL entity with functions *fifo_read* and *fifo_write* that are instances of READPORT and WRITEPORT objects. The *fifo* implements the interface *intf* that is described as an instance of an INTERFACE object. The connections c_1 and c_2 are instances of the CHANNEL2INTERFACE relationship that represents the implementation of the virtual read-write functions of the *intf* by the *fifo*. The textual description of the visual model of the FIFO is shown in Figure 19.8.a.

Another integral part of the metamodel is the constraints specified using OCL. Constraints are specified on entities (E_C) as well as on relationships (R_C), which are activated during instantiation/characterization in the case of an entity or during connection in the case of a relation. The activated constraints are checked during modeling to enforce conformance to the metamodel and are collected through I_C and C_C. We discuss a few constraints on an entity and relation for the purpose of illustration. The activation of *leaf_abstraction*

f_{id}(CHANNEL, fifo)

⋮

f_{id}(fifo.READPORT, fifo_read) ...
f_{id}(fifo.WRITEPORT, fifo_write) ...
f_{id}(fifo.INTERFACE, intf)
f_{id}(intf.WRITEPORT, write) ...
f_{id}(intf.READPORT, read) ...

⋮

f_{id}(CHANNEL2INTERFACE, c_1);
f_{id}(CHANNEL2INTERFACE, c_2);
c_1(fifo.fifo_read, intf.read);
c_2(fifo.fifo_write, intf.write);

a) Textual description of FIFO

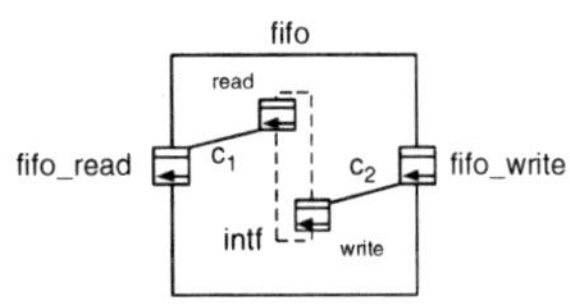

b) Block Diagram of FIFO

Figure 19.8. Fifo Channel implementing an interface.

constraint on a leaf instance (l) checks the following: (i) if d is RTL, then l is only allowed to contain IO ports, (ii) if d is TL, then l is only allowed to contain read-write ports, and (iii) l is not allowed to have a mixed RTL-TL description. Consider a connection $c \in C$ of type LEAF2LEAF that connects two LEAF instances namely l_1 and l_2 and therefore activates the *oport_to_iport* constraint when it connects the output port p_o of l_1 to the input port p_i of l_2. The *oport_to_iport* constraint checks whether the connected ports match in their datatype through type-equivalence or are unspecified and is formalized as shown below:

$$p_o.dt = p_i.dt \text{ or } p_o.dt = UN \text{ or } p_i.dt = UN$$

5. CC Analysis and Translation

Developing a CCL on top of a UML-based visual language brings in the advantages associated with visualization and object-oriented features. However, the flip-side is that the static requirements need to be explicitly formalized for such a framework. Some of these requirements are formalized and enforced using OCL and some require analysis of the complete model through inference and propagation. We enforce these through checkers. We have implemented three different checkers, namely a consistency checker that disallows any discrepancies in the CCM. The type propagation engine performs type assignment through type inference and type checking. A type checker looks for type inconsistencies through inference and propagation. Finally, a translator that converts the visual CCM into an intermediate representation that will serve as input to IComp.

Consistency Checking

Some of the HDL related consistency checks performed are whether connected components match in terms of port sizes, interfaces calls, whether unbounded ports or duplicate ports exist, etc. In this respect, the checker is similar to a linting tool for an HDL language. However, the components are devoid of an implementation, as a result, behavioral syntax and semantic checks cannot be performed by this checker. On the other hand, automatic selection of an implementation from the IP library requires the component to be annotated with sufficient metadata. For example, if the user wants to perform selection on a name basis, then the checker will enforce that the components have a unique identity. If two components have the same identity, then they are inferred as instances of the same component and checked for identical signatures. A consistency checker is an important part of the description and selection process since it finds inconsistency early in the specification that reduces design time.

Type checking and Propagation

Systems described with a strong-typed language such as C++ requires components to be implemented using interfaces that are type compatible. The programming effort associated with manually ensuring such type compatibility results in significant overhead that hinders the design process. A component's interface description using our CCMM is dependent on the level of abstraction and so are our checks. At the RTL-level, type-compatibility boils down to checking whether output and input ports of the communicating components match in terms of datatype. At the TL, type checking resorts to verifying that the read and write calls match in terms of function signatures (return values, argument types, and number of arguments).

We perform type checking at two levels firstly at design time using OCL constraints and during CC analysis. Some of the type checks are expressed as OCL constraints on the connections, which at the RTL enforces that ports of connected components are type equivalent (Definition 3.2). Figure 19.9 illustrates an RTL component C_1 connected to C_2 in Ex1 and to C_3 and C_4 in Ex2. if C_1 is fully typed with a 8-bit integer OUT port of size 2 in Ex1 then the constraints enforced on C_2 due to their connection are as follows:

$$C_2.IN.dt = C_1.OUT.dt \text{ and } C_2.IN.sz = C_1.OUT.sz$$

However in Ex2, if C_1 has a partially typed port of size 2 and C_3, C_4 are fully typed with inputs of size 1 and unrelated types, then their connections to C_1 will flag a violation. The type checker cannot infer a consistent type for the OUT port of C_1; therefore the OCL constraints will catch these types of errors at design time rather than having the type checker catch it later during analysis. The type checking at the TL resorts to verifying that function signatures match.

In the second phase, our checker is given a type-structure (Definition 3.1) and it achieves type propagation through inference algorithms and checking techniques. If the designer wants to use other type relations besides equivalence, then the constraints serve as notification that the user has connected two ports

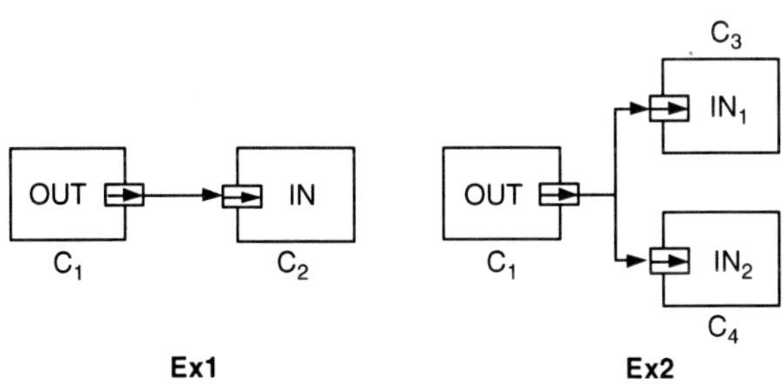

Figure 19.9. Abstract examples.

that are not type equivalent. Consider a SystemC example, if an adder's output of type *sc_int*<16> and the input of a multiplier of type *sc_bigint*<16> are connected, then the type checking will not flag an error, since they are related through type subsumption. In order, to allow type checking based on subsumption, we require a predefined type-structure, that the checker consults similar to Ptolemy's type system [12]. The checking based on type subsumption (Definition 3.3) is performed by our type checker that derives the type relations given a type-structure. Upon completion of type checking, the engine propagates the inferred type. One of the objectives of MCF is to promote reusability through abstract specification and provide an easy and fast specification. If the user wants to describe a generic adder, then the component is not bounded to a type-contract. Having a component conforming to an interface for manipulation and integration, restrict its reusability as it becomes dependent on the context of usage. Therefore, we allow partial specification, where a component does not have a fully typed interface. Consider the generic adder shown in Fig. 19.10, based on whether it is connected to two floating-point multipliers or two 16-bit integer adders, the type-contract derived are different.

Another reason for allowing partial-typed components is to avoid the tediousness associated with characterizing every component in the CCM. The types for some of the components can be inferred based on their static requirements or from their associations. For example, our CCMM provides communication media such as splitter and merger for muxing and demuxing. Their input-output types are inferred based on the component's output being (de-)multiplexed. Furthermore, inferring the type of one port of the splitter/merger would require that the other ports adhere to it through equivalence or subsumption. Therefore, for such media a single-port inference is used to propagate types for all the other partially typed (de-)multiplexing ports associated with it.

Similarly, another place where our type propagation comes handy is when dealing with hierarchical ports that serve as transfer ports between their inter-

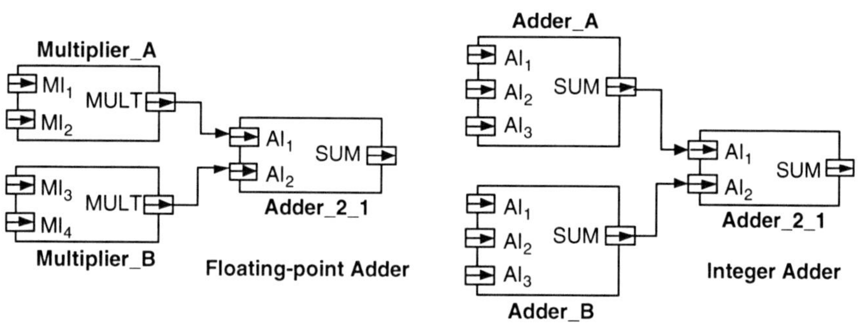

Figure 19.10. A generic Adder_2_1.

nals and externals. The details of the type inference algorithm for different components and medium have been provided in [1]. Our type propagation engine does a type inference and follows that up with a type assignment to propagate the inferred type. The type assignment necessitates another round of type checking to look for inconsistency that might have risen. These are type conflicts that crop up when the propagation engine tries to assign a type to component/media's internals which violates its type-contract. This conflict indicates that there exists a component/media that is not correctly specified or that participates in an illegal composition. Therefore a fix-pointing takes place when the type propagation tries to fully type the CCM.

Another reason for allowing partial typing and failure of type propagation to provide a type assignment is to allow for polymorphic components in the CCM. The modeler uses the CCMM for modeling a generic template of the architecture and various IP components can be used to instantiate different simulation models. This flexibility allows an automatic exploration of alternatives that best suits a CCM from a library of IPs. In C++ based HDLs such as SystemC, a component built using a C++ template are polymorphic and use type variables that are place holders for concrete types and resolved during instantiation. Some examples of such components/medium are buffers, fifos, memories and switches. Consider the *Adder_2_1* in Fig. 19.10 that is partially typed. Depending on whether *Adder_2_1* is connected to floating point multipliers or integer adders, the type specification and the instantiated implementation from the IP library differ.

XML Translation

The translator generates an XML representation (XML-IR) for the CCM description. This representation is validated against a Document Type Definition (DTD) for well-formed-ness. The translator uses a templated code generator that when given a component or media, prints out its corresponding XML representation. Upon completion, an XML graph is generated that describes their connectivity. The XML generation serves as input to the next stage of IP selection process.

The DTD for a LEAF component is shown in Listing 5.0.

```
// LEAF entity
<!ELEMENT Leaf (Inport | Output | InputClockPort | OutputClockPort | Readports
| Writeports )* >
<!ATTLIST Leaf name CDATA #REQUIRED>

<!ELEMENT InputPort EMPTY >
<!ATTLIST InputPort name CDATA #REQUIRED datatype CDATA #REQUIRED size
CDATA #REQUIRED >

<!ELEMENT OutputPort EMPTY >
<!ATTLIST OutputPort name CDATA #REQUIRED
datatype CDATA #REQUIRED size CDATA #REQUIRED >

```

```
<!ELEMENT Readports (Parameter)* >
<!ATTLIST Readports name CDATA #REQUIRED >

<!ELEMENT Writeports (Parameter)* >
<!ATTLIST Writeports name CDATA #REQUIRED >

<!ELEMENT Parameter EMPTY >
<!ATTLIST Parameter name CDATA #REQUIRED datatype CDATA #IMPLIED
IOtype CDATA #REQUIRED >

<!ELEMENT InputClockPort EMPTY >
<!ATTLIST InputClockPort name CDATA #REQUIRED datatype CDATA #REQUIRED >

<!ELEMENT OutputClockPort EMPTY >
<!ATTLIST OutputClockPort name CDATA #REQUIRED datatype CDATA #REQUIRED >
```

6. Conclusion and Future Work

Our experience from using MCF is summarized as follows: (i) drag-and-drop environment with minimal effort in designing systems; (ii) fast design exploration; (iii) easy to model at different abstractions as well as at mixed levels; (iv) natural way to describe hierarchical systems; (v) incremental design process through rigorous design-time checks; and (vi) highly extensible using GME for future requirements. As future work, we would equip our CCL with the Open Core Protocol (OCP) [7] to express a large variety of hardware communication behaviors that promotes reusability at different abstraction levels within RTL and TL. Furthermore, we will implement the automated reflective mechanism and the IComp to facilitate a full fledge CCF. The reflective mechanism would include extracting both structural and behavioral information from all the IPs in the library and constructing a data structure containing these meta-information. We will use a SystemC XML parser to annotate the SystemC-based IP with XML tags and a XML parser to extract the required information and populate our data structure. This data structure and the IR of the CCM from our visual front end will be given as input to our IComp to allow automatic IPs selection on two accounts, firstly structural versus behavioral and secondly flattened versus hierarchical.

Final goal An MCF modeler creates an abstract architectural specification of the system that is independent of any IP implementation specifics and then through a few user initiated actions on the GUI, creates executable models. The task of the MCF besides enabling a specification is to take the architectural template of the system and automatically plug in matching IPs from a library through type inference. The type inference facilitates automatic selection of matching IPs for the abstract components in the architectural template and composes them to create simulation models for the design alternatives. The advantage is the reduction in design time and tediousness associated with constructing high-level models for design space exploration through the automated tool support.

Notes

1. CCM = CCM("DesignView") for complete design visualization

References

[1] Mathaikutty, D., and Shukla S. (2005). A metamodel for component composition. Technical Report 2005-13, Virginia Tech.

[2] Doucet, F., Shukla, S., Otsuka, M., and Gupta, R. (2002). An environment for dynamic component composition for Efficient co-design. In: *proc. Design Automation and Test Conference Europe*.

[3] Doucet, F., Shukla, S., Otsuka, M., and Gupta, R. (2003). Balboa: a component based design environment for system models. *IEEE Transactions on Computer Aided Design*, 22(12):1597–1612.

[4] Explorations, Y. http://www.yxi.com.

[5] Ledeczi, A., Maroti, M., Bakay, A., and Karsai, G. (2001). The generic modeling environment. In: *Institute for Software Integrated Systems Nashville*. Vanderbilt University.

[6] Lennard, C. K., and Granata, E. (1999). The meta-Methods: managing design risk during IP selection and integration. In: *Proceedings of European IP 99 Conference*.

[7] OCP OCP International Partnership Association, Inc. www.OCPIP.org.

[8] SPIRIT Spirit Consortium, SPIRIT Schema Working Group. http://www.spiritconsortium.org/.

[9] The Metropolis Project Team (2004). The metropolis meta model version 0.4. UCB/ERL M04/38, Department of Electrical Engineering and Computer Science, University of California, Berkeley.

[10] Vachharajani, M., Vachharajani, N., Malik, S., and August, D. (2004). Facilitating reuse in hardware models with enhanced type inference. In: *Proc*. The IEEE/ACM/IFIP Second International Conference on Hardware/Software Codesign and System Synthesis, pp. 86–91.

[11] Vachharajani, M., Vachharajani, N., Penry, D., Blome, J., and August, D. (2002). Microarchitectural Exploration with Liberty. In: *proc. of the 35th International Symposium on Microarchitecture (MICRO)*, November.

[12] Xiong, Y. (2002). An extensible type system for component based design. Ph.D thesis, University of California, Berkeley, Electrical Engineering and Computer Sciences.

Chapter 20

REUSING SYSTEMS DESIGN EXPERIENCE THROUGH MODELLING PATTERNS*

Oana Florescu[1], Jeroen Voeten[1,2], Marcel Verhoef[3], and Henk Corporaal[1]

[1]*Eindhoven University of Technology,*
P.O. Box 513, 5600 MB Eindhoven,
The Netherlands
o.florescu@tue.nl
h.corporaal@tue.nl

[2]*Embedded Systems Institute,*
P.O. Box 513, 5600 MB Eindhoven,
The Netherlands
j.p.m.voeten@tue.nl

[3]*Chess Information Technology B.V.,*
P.O. Box 5021, 2000 CA Haarlem,
The Netherlands
Marcel.Verhoef@chess.nl

Abstract Based on design experience for real-time systems, we introduce modelling patterns to enable easy composition of models for design space exploration. Our proposed approach does not require deep knowledge of the modelling language used for the actual specification of the model and its related analysis techniques. The patterns proposed in this paper cover different types of real-time tasks, resources and mappings, and include also aspects that are usually ignored in classical analysis approaches, like task activation latency or execution context switches. In this paper, we present a library of such modelling patterns expressed in the POOSL language. By identifying the patterns that are needed for the specification of a

*This work is being carried out as part of the Boderc project under the responsibility of the Embedded Systems Institute. This project is partially supported by the Netherlands Ministry of Economic Affairs under the Senter TS program.

S.A. Huss (ed.), Advances in Design and Specification Languages for Embedded Systems – Selected Contributions from FDL'06, 339–358.

system, the POOSL model can be automatically generated and analysed using its related tools and techniques as illustrated in two case studies.

Keywords Real-time systems, modelling patterns, design space exploration models

1. Introduction

Complex real-time embedded systems are usually comprised of a combination of hardware and software components that are supposed to synchronise and coordinate different processes and activities. From early stages of the design, many decisions must be made to guarantee that the realisation of such a complex machine meets all the functional and non-functional (timing) requirements.

One of the main problems to address concerns the choice of the most suitable architecture of the system such that all the requirements are met. To properly deal with this question, the common approaches are design space exploration and system-level performance analysis. An extensive overview of such methodologies is given in [2] and [14]. They range from analytical computation to simulation-based estimation. These are oft-specialised techniques which claim that general purpose languages are ill-suited for system-level analysis. However, due to the heterogeneity and complexity of systems, for the analysis of different aspects different models need to be built and their coupling is often difficult. Therefore, a unified model, covering all the interesting aspects, is actually needed to speed-up the design process. This is how the Unified Modelling Language (UML) [20] came to be conceived. The language was designed mainly for object-oriented software specification, but recently it was extended (UML 2.0) to include (real-time) systems as well.

During the development of new systems, common problems are encountered again and again, and experienced designers apply the solutions that worked for them in the past [11]. These pairs of problem-solution are called *design patterns* and their application helps in getting a design "right" faster. With the increase in the development of real-time systems, design patterns were needed for dealing with issues like concurrency, resource sharing, distribution [6]. As UML has become the industry standard language for modelling, these patterns are described in UML. However, the semantics of the language is not strong enough to properly deal with the analysis of *real-time* system behaviour. Therefore, an expressive and formal modelling language is required instead in order to capture in a compact model *timing*, *concurrency*, *probabilities* and *complex behaviour*.

Design patterns refer to problems encountered in the design process itself, but problems appear also in the specification of components that are commonly encountered in complex systems [13]. Although components of the analysed systems exhibit some common aspects for all real-time systems (e.g. characteristics of tasks like periodicity or aperiodicity, processors, schedulers, and their overheads), they are built everytime from scratch and similar issues are encountered over and over.

Contributions of the paper. To reduce the amount of time needed to construct models for design space exploration, we propose *modelling patterns* to easily compose models for the design space exploration of real-time embedded systems. These modelling patterns, provided as a library, act like templates that can be applied in many different situations by setting the appropriate parameters. They are based on the concepts of a mathematically defined general-purpose modelling language, POOSL [25], and they are presented as UML diagrams. These boilerplate solutions are a critical success factor for the practical application in an industrial setting and are a step towards the (semi-) automated design space exploration in the early phases of the system life cycle.

This paper is organised as follows. Section 2 discusses related research work. Section 3 briefly presents the POOSL language, whereas Section 4 provides the modelling patterns. The composition of these patterns into a model is discussed in Section 5 and their analysis approach in Section 6. The results of applying this approach on two case studies are presented in Section 7. Conclusions are drawn in Section 8.

2. Related Research

An extensive overview of performance modelling and analysis methodologies is given in [2] and [14]. They range from analytical computation, like Modular Performance Analysis [26] and UPPAAL [3], to simulation-based estimation, such as Spade [18] or Artemis [21]. The techniques for analytically computing the performance of a system are exhaustive in the sense that *all* possible behaviours of the system are taken into account. For this reason, the models are built at a high level of abstraction and they abstract away from the non-deterministic aspects, otherwise the state space explosion problem is faced.

On the other hand, simulation of models allows the investigation of a *limited* number of all the possible behaviours of the system. Thus, the obtained analysis results are *estimates* of the real performance of the system. To obtain credible results, these techniques require the models created to be amenable to mathematical analysis (see [23]), using mathematical structures like Real-Time Calculus [5], timed automata [1] or Kahn process networks [16]. As in general, analytical approaches do not scale with the complexity of the industrial systems, simulation-based estimation of performance properties is used more often. In this context, the estimation of performance is based on statistical analysis of simulation results.

The analysis technique described in this paper is based on simulation of models expressed in the formally defined modelling language POOSL. Due to the semantics of the language, analytical computation of the properties of a real-time system *is possible*. However, the type of models that we shall present throughout the paper, though compact, they incorporate non-determinism. Due to the state space explosion problem, estimation of the system performance is used instead of analytical computation.

With respect to the idea of using patterns for building a model based on previous experience with modelling systems from a particular application area, this is not new. As an example, in [13], the authors propose patterns to deal with the complexity of system models by reusing structures expressing expert modelling experience at a higher level of design and abstraction than the basic elements. In a similar manner, we propose modelling patterns for real-time systems that capture their typical components and characteristics, like tasks, computation and communication resources, schedulers, and input and output devices. The use of them prevents the building of system models from scratch over and over and enable the possibility of automatic generation of such models based on a textual representation.

3. POOSL Modelling Language

The Parallel Object-Oriented Specification Language [25] lies at the core of the Software/Hardware Engineering (SHE) system-level design method. POOSL contains a set of powerful primitives to formally describe concurrency, distribution, synchronous communication, timing, and functional features of a system into a single executable model. Its formal semantics is based on timed probabilistic labelled transition systems [22]. This mathematical structure guarantees a unique and an unambiguous interpretation of POOSL models. Hence, POOSL is suitable for specification and subsequently, verification of correctness and evaluation of performance for real-time systems.

POOSL consists of a *process* part and a *data* part. The process part is used to specify the behaviour of active components in the system, the processes, and it is based on a real-time extension of the Calculus of Communicating Systems [19]. The data part is based on traditional concepts of sequential object-oriented programming. It is used to specify the information that is generated, exchanged, interpreted, or modified by the active components. As mostly POOSL processes are presented in this paper, Fig. 20.1 presents the relation between the UML class diagram and the POOSL process class specification. The name compartment of the class symbol for process classes is stereotyped with <<process>>. The attributes are named <<parameters>> and allow parameterising the behaviour of a process at instantiation. The behaviour of a process is described by its <<methods>> which may include the specification of sending (!) and/or

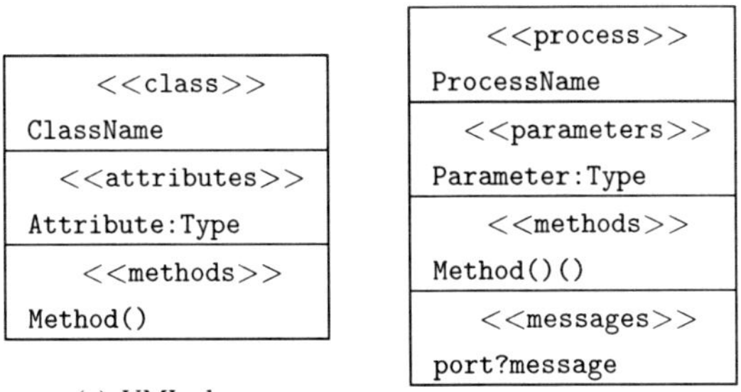

(a) UML class

(b) POOSL process class

Figure 20.1. UML vs. POOSL process class.

receiving (?) of `<<messages>>`. More details about the UML profile for POOSL can be found in [23].

The SHE method is accompanied by two tools, SHESim and Rotalumis. SHESim is a graphical environment intended for incremental specification, modification and validation of POOSL models. Rotalumis is a high-speed execution engine, enabling fast evaluation of system properties. Compared with SHESim, Rotalumis improves the execution speed by a factor of 100. Both tools have been proved to correctly simulate a model with respect to the formal semantics of the language in [12].

4. Modelling Patterns

One of the approaches for performing systematic design space exploration is the Y-chart scheme, introduced in [17]. This scheme makes a distinction between applications (the required functional behaviour) and platforms (the infrastructure used to perform this functional behaviour). Although we are concerned only with the realisation of the software part of a real-time system, the hardware part must also be taken into account in the analysis in order to predict the behaviour of the system as a whole and the impact each part may have on the others. Moreover, as real-time systems are typically reactive systems, meaning that there is a continuous interaction with the outside world, in [10] we added the model of the environment to the Y-chart scheme, as depicted in Fig. 20.2. The design space can be explored by evaluating different mappings of applications onto platforms.

To reduce the amount of time needed to construct models for design space exploration, we propose *modelling patterns* to easily compose models of real-time embedded systems. They are provided as a library and act like templates that can be applied in many different situations by setting the appropriate parameters. These modelling patterns emerged from previous experience with real-time systems modelling (see [9], [7], and [24]). Moreover, they reflect

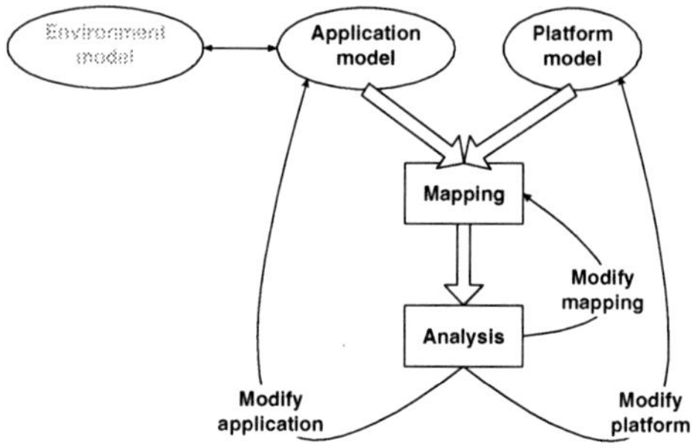

Figure 20.2. Y-chart scheme.

Y-chart Part	Pattern Name	Parameter Names	
Application Model	PeriodicTask	period (T) deadline (D) load	latency (l) iterations
	AperiodicTask	deadline (D) load	latency (l)
Platform Model	Resource	initial latency	throughput
	Scheduler	scheduling policy	
Environment Model	InputGenerator	event type	generation stream
	OutputCollector	event type	desired throughput

Figure 20.3. Modelling patterns.

the approach assumed in classical scheduling analysis [4] for modelling such systems. The library of patterns contains templates for different types of tasks, resources, schedulers, and input and output devices, which are presented in Fig. 20.3. This figure shows the Y-chart components to which each of these patterns belongs, the names of the patterns and their parameters. Each of these patterns are explained in the remainder of this section.

Application Model

The functional behaviour of a real-time embedded system is implemented through a number of tasks that may communicate with each other. Task activation requests can be periodic (time-driven), being activated at regular intervals equal to the task period T, or aperiodic (event-driven), waiting for the occurrence of a certain event. There are three types of uncertainties, showed in Fig. 20.4, that may affect a task:

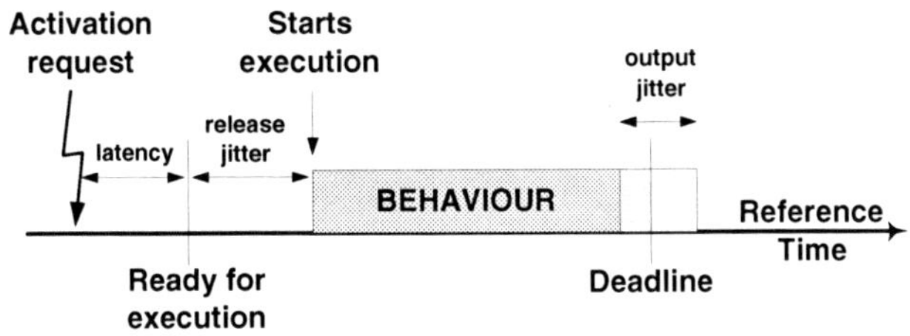

Figure 20.4. Real-time task parameters.

- Activation latency: It may be caused, for example, by the inaccuracies of the processor clock that might drift from the reference time because of temperature variations. For event-driven tasks, the performance of the run-time system, which cannot continuously monitor the environment for events, may also have influence.

- Release jitter: It may be caused by the interference of other tasks that, depending on the scheduling mechanism, may impede the newly activated task to start immediately its execution.

- Output jitter: It may be caused by the cumulated interference of other tasks in the system, the scheduling mechanism that may allow preemption of the executing task, the variation of the activation latency, and even of the execution time of the task itself, which may depend on the input data.

In classical real-time scheduling theory [4], the release jitter and, to some extent, the output jitter can be computed, but the activation latency is usually ignored. As in control-oriented systems the effect of this latency might be of significant importance, the POOSL specification of the task modelling patterns overcomes this problem. The two task patterns that we have conceived are visualised using the POOSL equivalent of UML class diagrams in Fig. 20.5.

The `PeriodicTask` pattern is to be used whenever a periodic independent task is required. Its parameters are the period `T`, the deadline `D`, the `load`, which represents the worst-case value of the number of instructions the task imposes on a target processor and can be obtained based on previous experience, the activation latency `l` specified as a distribution, and the number of `iterations` for the case the task is not infinitely running. The `AperiodicTask` pattern should be applied for the specification of a task triggered by an event from the environment or by a message from another task in the system. Its parameters are the deadline `D`, the `load`, and the activation latency `l`. Each of these two patterns has two methods. One is called `Periodic`, and respectively `Aperiodic`, and contains the specification of the task according to the type of triggering it has,

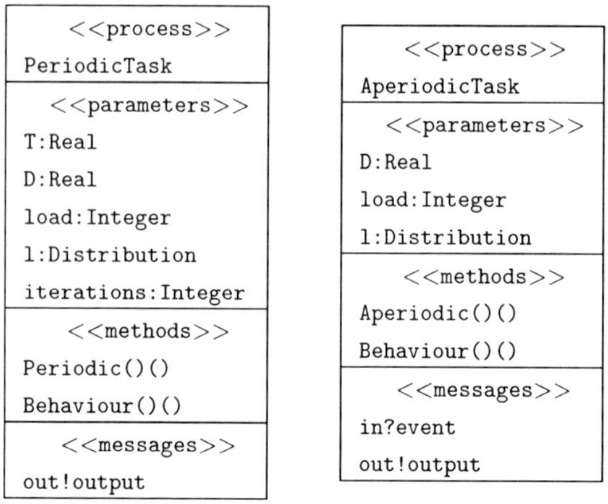

(a) Periodic task pattern

(b) Aperiodic task pattern

Figure 20.5. Application model patterns.

such as waiting for the next period, respectively for the next incoming event, to be activated, whereas the `Behaviour` method contains the specification of the actual computation the task needs to perform. In the templates provided with our library, the specification of this method is empty for two reasons. The first one is that it depends on the application what a task is supposed to compute, hence the designer who is using this library has to supply the right specification. The second reason is that for the type of analysis we are interested at a high level of abstraction, which will be discussed in Section 6, the actual computation of a task is not important and can be left out.

Platform Model

The modelling patterns we have conceived for describing the platform part of the Y-chart model of a system provide a unified way of specifying communication and computation resources by exploiting their common characteristics. This modelling approach is possible as at a high level of abstraction there is no large conceptual difference between a processor and a bus: they both receive requests, execute them (either by transferring the bits of a message or executing the instructions of an algorithm) and send back a notification on completion. As a resource is typically shared, a scheduler is needed in order to arbitrate the access to a resource.

Figure 20.6 visualises the `Scheduler` and the `Resource` modelling patterns that are needed for the specification of the platform model. The scheduler has one parameter which is the name of the scheduling policy desired to be used on a certain resource. Amongst the scheduling policies that we provide within

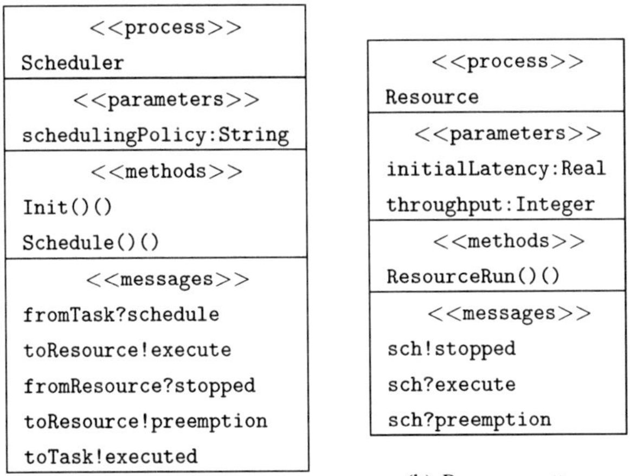

(a) Scheduler pattern

(b) Resource pattern

Figure 20.6. Platform model patterns.

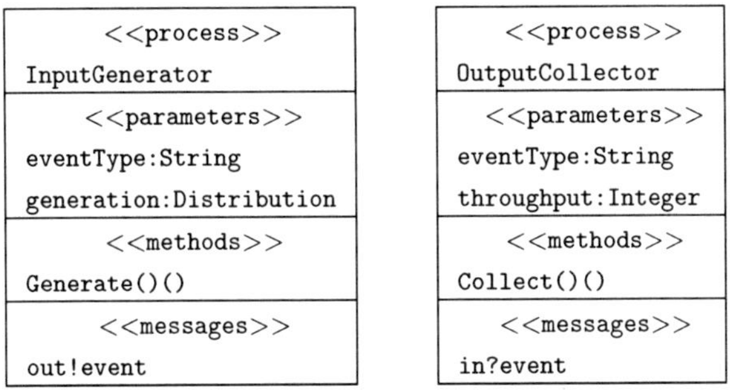

(a) Input generator pattern

(b) Output collector patterns

Figure 20.7. Environment model patterns.

the POOSL library are: earliest deadline first, rate monotonic, first come first served, and round-robin. The resource is characterised by a `throughput`, which is the number of instructions a processor can execute per time unit or the transfer bit rate on a bus, and an initial latency, which incorporates the task context switch time or the communication protocol overhead.

Environment Model

The model of the environment is composed by input generators and output collectors, as showed in Fig. 20.7. The input generators model the generation of environmental events of a certain `eventType` with a certain `generation` pattern which can be chosen amongst periodic, periodic with jitter, or sporadic with a certain distribution of occurrence, such as uniform or normal. These events trigger the activation of tasks in the application model and are collected by output collectors which model the output devices in the environment. The collector receives the events of a certain `eventType` exiting the system and compares the end-to-end delay of the system against the desired one expressed as the `throughput`.

5. Model Composition

To build a model of a real-time system for design space exploration, its specific components that correspond to the modelling patterns described in the previous section must be identified together with their parameters. The names of the necessary patterns and their parameters, together with the specification of the mapping (which task is scheduled on which processor, etc.) and the layout of the platform (which processor is connected to which bus) can be provided as the configuration of the system. From such a configuration, the POOSL model of the system can be automatically generated based on the library of modelling patterns and fed to SHESim or Rotalumis tools for analysis.

As an example, a producer-consumer system is showed in Fig. 20.8(a). The system is made of a periodic task producer, `TASK1`, and an aperiodic task consumer, `TASK2`, whose activation is triggered by the production of a new item by `TASK1`. The specification of the system may look like the one in Fig. 20.8(b) structured along the Y-chart scheme, expressing the application components, the platform, and its interconnections, and the mapping. The structure of the generated model is showed in Fig. 20.8(c). As it can be seen, it differs somewhat

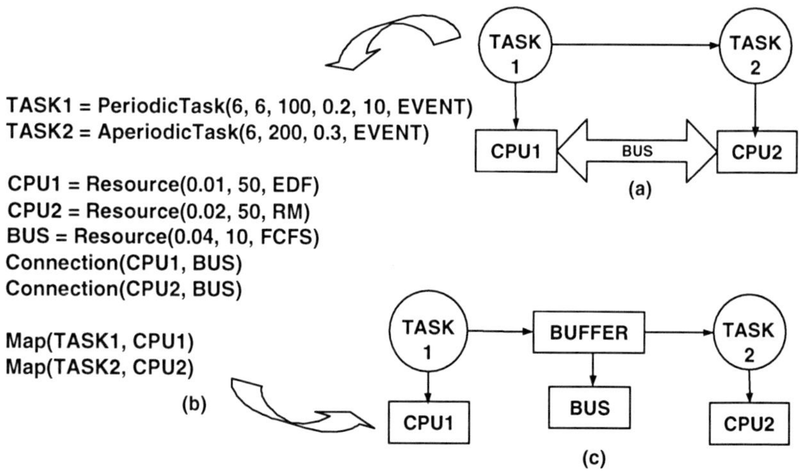

Figure 20.8. Model specification based on modelling patterns.

from the original system in the sense that the model generation tool is able to detect that the communication between the two tasks is done over a bus, hence a buffer to contain the message and to transport it across the communication medium is required.

For design space exploration, different configurations must be compared. To do this, changes in the initial configuration may be done and the POOSL model regenerated in order to analyse them. To specify a different mapping, the `Map` specifications must be changed according to the new task-to-resource mapping. To change the architecture components, simply change the `Resource` specifications and/or their parameters. Similarly, the interconnections of the platform can be changed in the `Connection` specification tags. In this way, the model can be easily tuned to specify different possibilities in the design space without any knowledge about the underlying formal model that will be generated in accordance with the description of the new configuration.

6. Model Analysis

By composing together the necessary modelling patterns, the complete model of a system can be built and validated. For each configuration specified and generated, during the execution of the model, the scheduler, which also acts as a monitor for the system schedulability, can report if there are any tasks that miss their deadlines. Furthermore, based on the formal semantics of POOSL, it can be analysed if there is any deadlock in the system. If all the deadlines are met and there is no deadlock, then the corresponding architecture is a good candidate that meets the system requirements.

However, for soft real-time systems, it is allowed that some deadlines are missed (usually there is a given threshold). Therefore, in this case, it is especially useful that in the specification of the tasks their computations are decoupled from the activation mechanism, in the sense that the analysis of the model could handle tasks with multiple active instantiations that are likely to miss their deadlines. The percentage of deadlines missed can be monitored and checked against the requirements if, according to this criterion, the underlying platform is suitable.

To correctly dimension a system (the required CPUs and BUSes performance) such that it works in any situation, the worst-case behaviour of the system must be analysed. This usually means to consider the worst-case execution times for all the activities in the system. On the other hand, the analysis of the average behaviour, based on probabilities, which can be enabled in the proposed patterns, as showed in [8], gives a measure of the suitability of the design. If the dimension of the system, needed for the worst-case situation that appears only once in a while, is far bigger than the one needed in average, that could give useful hints for a redesign (e.g. split tasks into smaller ones in order to spread the load onto different CPUs).

Some other useful results the analysis of the proposed model can provide are the release and the output jitter of tasks, which is useful for control applications, the number of active instances of a task type, which influences the missed of deadlines, the throughput of the system, important in streaming applications.

7. Case Studies

In this section, two case studies are presented for which worst-case analysis and design space exploration have been performed using the modelling patterns proposed in this work. The characteristics of the systems and the results of their analysis follow.

A Printer Paper-Path

The first case study is inspired by a system architecture exploration for the control of the paper-path of a printer.

The high-level view of the system model, visualised using SHESim tool, is given in Fig. 20.9. User's printing requests arrive at the high-level control (HLC) of the machine which computes which activities need to take place and when in order to accomplish the request. The HLC tasks activate the tasks representing the low-level control (LLC) of the physical components of the paper path, like motors, sensors, and actuators. As HLC tasks are soft real-time, whereas LLC tasks (Fig. 20.10) are hard real-time, a rather natural solution was to consider a distributed architecture. LLC can be assigned to dedicated processor(s) and connected through a network to the general-purpose processor that runs HLC.

Under these circumstances, the *problem* was mainly *to find an economical architecture for LLC*, whose task parameters are showed in Fig. 20.11. For the models of the time-driven tasks of type T1, T3, and T4, we took into account a latency of up to 10% of their period. Although tasks of type T2 are activated

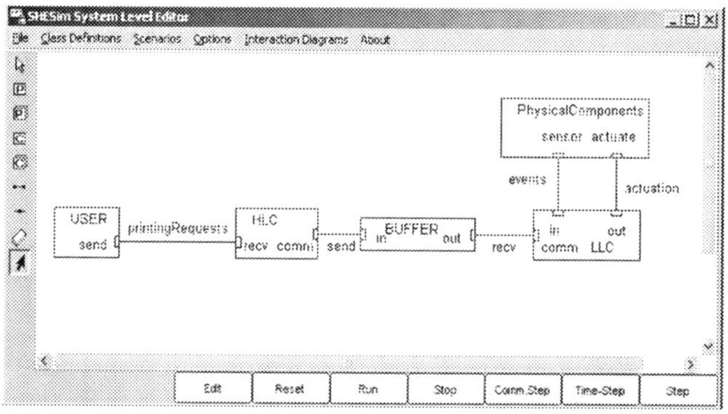

Figure 20.9. High-level printer control POOSL model.

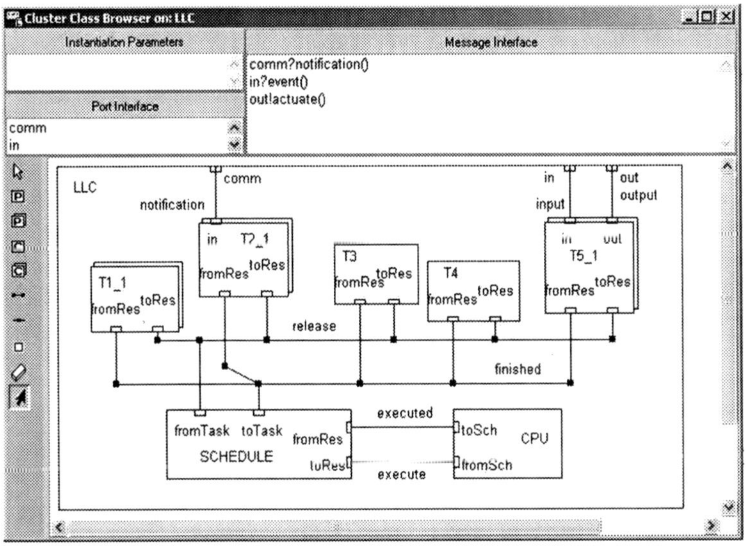

Figure 20.10. POOSL low-level control model.

Task type	No. of instantiations	Load [*instr.*]	T [*ms*]	D [*ms*]
T1	3	3200	2	2
T2	8	1200	2	2
T3	1	2000	2	2
T4	3	800	0.66	0.1
T5	4	160	–	0.064

Figure 20.11. Low-level control task parameters.

based on notifications from HLC, they behave completely periodic until the next notification arrives. Therefore, their dynamical behaviour was captured using an aperiodic task which triggers a periodic task with a finite number of activations. Tasks of type T5 are event-driven; therefore, a model of the environment was needed (*PhysicalComponents*), for which we considered event streams with a uniform distribution in $[1, 20]$ ms.

Given the frequency of events and the task execution times, we have analysed three commercially available low-end processors, a 40 MIPS, a 20 MIPS, and a 10 MIPS, and compared their utilisations under different schedulers. Figure 20.12 presents the results obtained using the earliest deadline first scheduling algorithm [4]. Although the 10 MIPS processor seems to be used the most efficiently (close to its maximum capacity), the analysis of the model

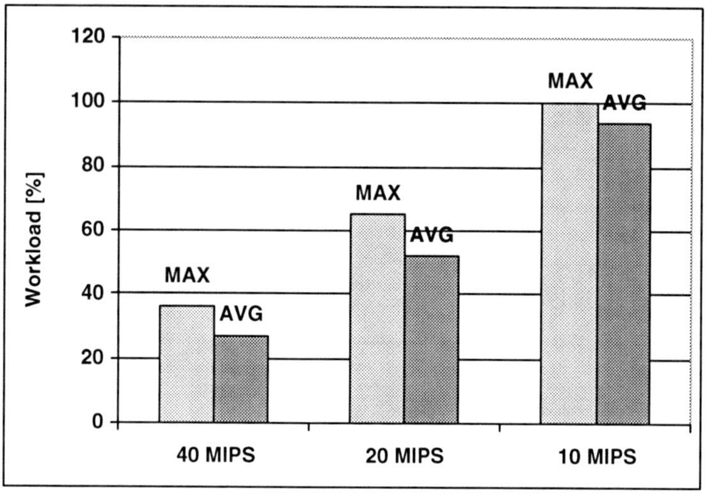

Figure 20.12. CPU workload comparison.

Task type	Release jitter [*ms*]	Output jitter [*ms*]
T1	0.466	1.852
T2	0.466	1.852
T3	0.414	1.884
T4	0.042	0.128
T5	0.472	1.094

Figure 20.13. Tasks jitter for the 20 MIPS.

showed that some of the deadlines are missed; thus this processor is not a good candidate. For the other two, all the deadlines are met. Due to the fast execution engine Rotalumis, tens of hours of system behaviour could be covered in less than one minute simulation. Moreover, the analysis of the model gave the values of the maximum release jitter, respectively output jitter of the tasks (for the 20 MIPS they are showed in Fig. 20.13) which could be checked against the expected margins of errors of the control design.

An In-Car Navigation System

The second case study is inspired by a distributed in-car navigation system described in detail in [26]. The high-level view of the system is presented in Fig. 20.14(a). There are three clusters of functionality, as the picture suggests: the man–machine interface (MMI) that handles the interaction with the user; the navigation functionality (NAV) that deals with route-planning and navigation

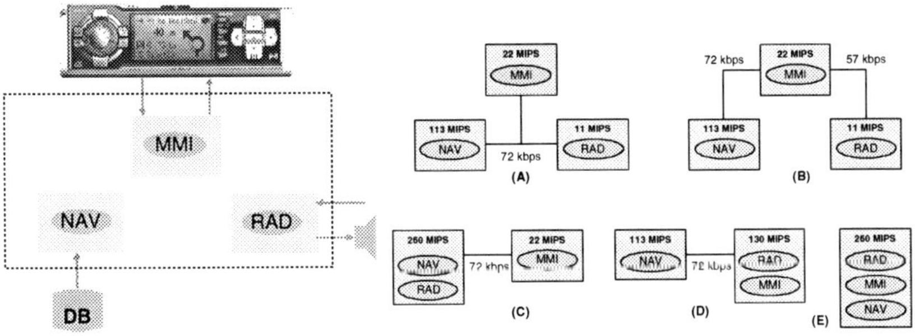

(a) The high-level view of the system (b) Platforms proposed for analysis

Figure 20.14. In-car radio navigation system.

Scenario	Task	Load [*instr.*]	T [*s*]	D [*s*]
I	T1	1E5	1/32	1/32
	T2	1E5	1/32	1/32
	T3	5E5	1/32	1/32
II	T4	1E5	1	1
	T5	5E6	1	1
	T6	5E5	1	1
III	T7	1E6	3	3
	T8	5E6	3	3
	T9	5E5	30	30

Figure 20.15. In-car navigation systems tasks.

guidance; the radio (RAD) which is responsible for basic tuner and volume control, as well as receiving traffic information from the network. Three application scenarios are possible. Users are allowed to change the volume (scenario I) and to look addresses up in the maps in order to plan their routes (scenario II); moreover, the system needs to handle the navigation messages received from the network (scenario III). Each of these scenarios has its own individual timeliness requirements that need to be satisfied. They all share the same platform, however, not all three of them can run in parallel due to the characteristics of the system (only I with III, or II with III). The characteristics of tasks for each scenario are given in Fig. 20.15. They are all periodic tasks with infinite behaviour. Notice that, in comparison with the previous case study, the timing requirements of this system are in the seconds domain and the loads imposed on the resources are much larger, as this case study combines control with data streaming.

The *problem* related to this system was *to find suitable platform candidates* that meet all the timing requirements of the application. For exploration of the design space, a few already available platforms (see Fig. 20.14(b)) were proposed for analysis. Two approaches have been applied for the analysis of this system, Modular Performance Analysis (MPA) in [26] and UPPAAL in [15].

MPA is an analytical technique in which the functionality of a system is characterised at a high level of abstraction by quantifying the incoming and outgoing event rates, message sizes, and execution times. Based on Real-Time Calculus, hard upper and lower bounds of the system performance are computed. Although these bounds are always hard, they are in general not tight, meaning that the technique derives conservative estimates of worst and best case.

The UPPAAL model checker is a tool for modelling and verifying networks of timed automata. The analysis results obtained by applying this technique are exact computations of the performance properties. Nevertheless, the method suffers severely from the state space explosion problem. Limitations, stating for example that tasks can be preempted only up to a certain number of times, are necessary, otherwise model checking is not possible anymore. Moreover, combination of scenarios with large difference in the time scale of the requirements (milliseconds versus seconds) proved to be another problem for the model checker.

In the rest of this section, we show that the model of the in-car navigation system, which can be easily built using the modelling patterns (see Fig. 20.16), can also be accurately analysed. All the tasks in the system are event-driven, hence we could easily use just the `AperiodicTask` pattern with the parameter values showed in Fig. 20.15 for the construction of the application model, whereas for the environment we assumed streams of events with periodic arrival patterns. The end-to-end delay for each scenario on each of the

proposed platforms, in the presence or absence of other scenarios, was monitored. The analysis shows that all the timing requirements are met for all scenarios in all configurations. As an example, the results obtained for the worst case end-to-end delay for different combinations of scenarios on architecture `A` are presented in Fig. 20.17 next to the results obtained using MPA and UPPAAL techniques. As MPA is an analysis technique which finds hard upper bounds, this explains why its results are larger than the other techniques. On the other hand, the results computed by UPPAAL are exact values of the worst case end-to-end delay. It is interesting to observe that our results are very close to UPPAAL ($\sim$1% difference which also represents the accuracy of the results), except for scenario III for which the difference is 7%. For this situation, there

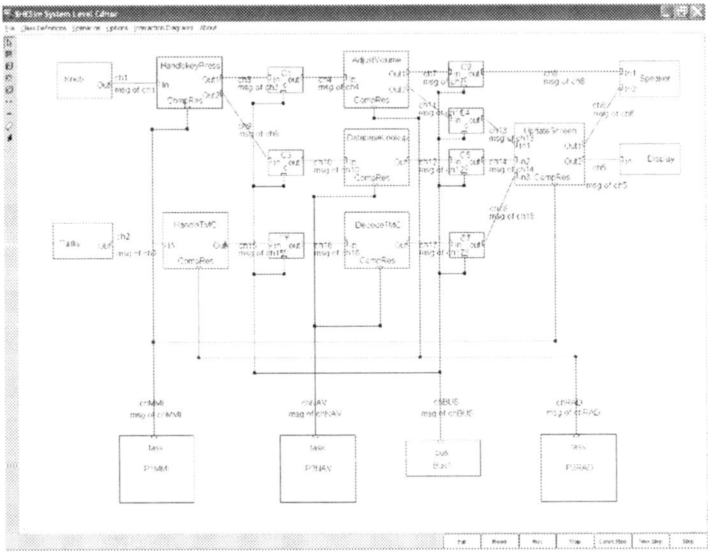

Figure 20.16. POOSL model of the in-car navigation system.

Measured scenario	Other active scenario	POOSL [*ms*]	MPA [*ms*]	UPPAAL [*ms*]
I	III	41.771	42.2424	41.796
III	I	357.81	390.086	381.632
II	III	78.89	84.066	79.075
III	II	171.77	265.849	172.106

Figure 20.17. Architecture `A` worst case end-to-end delays.

was a mismatch in the conceiving of the models with respect to the modelling of jitter in the incoming events.

Furthermore, the processor(s) and bus(es) utilisations were monitored and, as an example, Fig. 20.18 shows the results obtained for architecture `A`. All together, such results help the designer in detecting if there is any scenario likely to miss its deadline, or which processor or bus might be a bottleneck, and in choosing an appropriate platform.

Due to the easiness of using the patterns and going to different configurations in the design space by just changing their parameters, the construction of models for each of the proposed combinations took several minutes. Moreover, as mentioned for the previous case study as well, due to Rotalumis, the engine for the model execution, the analysis results could be obtained also fast.

Scen. I	Scen. II	Scen. III	MMI %	NAV %	RAD %	Bus %
YES	NO	NO	87	0	30	3
NO	YES	NO	3	5	0	1
NO	NO	YES	1	2	4	1
YES	NO	YES	88	2	33	4
NO	YES	YES	4	6	2	2

Figure 20.18. Processors and bus utilisations in architecture A.

8. Conclusions

In this paper, we have presented a library of modelling patterns, specified using the Parallel Object-Oriented Specification Language, that enables the automatic construction of models for the design space exploration of real-time embedded systems. To build such models, knowledge about the POOSL language itself is not needed as system models consisting of real-time tasks, computation and communication resources and their associated schedulers are specified in terms of the necessary patterns and the values of their parameters. Due to the expressiveness of POOSL, important aspects like task activation latencies and context switches can be taken into account, enabling the building of realistic models without sacrificing their conciseness. Moreover, due to this reason, the analysis can provide more realistic results than the classical scheduling techniques can.

The use of the patterns presented in this paper reduces both the modelling and the analysis effort. Although completeness cannot be claimed, the efficiency of the model simulation allows exploration of a substantial part of the design space. As future work, we aim at extending the modelling patterns to cover for complex platforms like networks-on-chip, by taking into account memory components, routing algorithms and even batteries for the analysis of energy consumption.

References

[1] Alur, R. and Dill, D.L. (1994). A theory of timed automata. *Theoretical Computer Science*, 126(2):183–235.

[2] Balsamo, S., Marco, A.D., Inverardi, P., and Simeoni, M. (2004). Model-based Performance Prediction in Software Development: A survey. *IEEE Transactions on Software Engineering*, 30(5):295–310.

[3] Behrmann, G., David, A., and Larsen, K.G. (2004). A Tutorial on UPPAAL. In: *Int. School on Formal Methods for the Design of Computer, Communication and Software Systems*, pp. 200–236.

[4] Buttazzo, G.C. (1997). *Hard Real-Time Computing Systems: Predictable Scheduling Algorithms and Applications.* Kluwer Academic Publishers, Boston, MA.

[5] Chakraborty, S., Kunzli, S., and Thiele, L. (2003). A general framework for analysing system properties in platform-based embedded system designs. In: *Proc. of Conference on Design, Automation and Test in Europe.* IEEE Computer Society.

[6] Douglass, B.P. (2002). *Real-Time Design Patterns: Robust Scalable Architecture for Real-Time Systems.* Addison-Wesley Longman Publishing Co., Inc.

[7] Florescu, O., de Hoon, M., Voeten, J., and Corporaal, H. (2006a). Performance modelling and analysis using poosl for an in-car navigation system. In: *Proc. the 12th Annual Conference of the Advanced School for Computing and Imaging*, pp. 37–45.

[8] Florescu, O., de Hoon, M., Voeten, J., and Corporaal, H. (2006b). Probabilistic modelling and evaluation of soft real-time embedded systems. In: *Proc. the Embedded Computer Systems: Architectures, Modeling, and Simulation*, LNCS 4017, pp. 206–215.

[9] Florescu, O., Voeten, J.P.M., and Corporaal, H. (2004a). A unified model for analysis of real-time properties. In: *Preliminary Proc. the 1st International Symposium on Leveraging Applications of Formal Methods*, pp. 220–227. TR-2004-6.

[10] Florescu, O., Voeten, J.P.M., Huang, and Corporaal, H. (2004b). Error estimation in model-driven development for real-time software. In: *Proc. Forum on Specification & Design Languages*, pp. 228–239. ISSN 1636-9874.

[11] Gamma, E., Helm, R., Johnson, R., and Vlissides, J. (1995). *Design Patterns: Elements of Reusable Object-oriented Software.*

[12] Geilen, M.G.W. (2002). *Formal Techniques for Verification of Complex Real-Time Systems.* Ph.D. thesis, Eindhoven University of Technology.

[13] Gries, M., Janneck, J., and Naedele, M. (1999). Reusing design experience for petri nets through patterns. In: *Proc. High Performance Computing*, pp. 453–458.

[14] Gries, M. (2004). Methods for evaluating and covering the design space during early design development. *Integration*, 38(2):131–183.

[15] Hendriks, M. and Verhoef, M. (2006). Timed automata based analysis of embedded system architectures. In *Workshop on Parallel and Distributed Real-Time Systems.*

[16] Kahn, G. (1974). The semantics of simple language for parallel programming. In: *Proc. IFIP Congress.*

[17] Kienhuis, B., Deprettere, E., Vissers, K., and van der Wolf, P. (1997). An approach for quantitative analysis of application-specific dataflow architectures. In: *Proc. the IEEE Int. Conference on Application Specific Systems, Architectures and Processors*, pp. 338–349. IEEE Computer Society.

[18] Lieverse, P., van der Wolf, P., Vissers, K., and Deprettere, E. (2001). A methodology for architecture exploration of heterogeneous signal processing systems. *VLSI Signal Processing Systems*, 29(3):197–207.

[19] Milner, R. (1989). *Communication and Concurrency*. Prentice-Hall.

[20] OMG (2003). *Unified Modeling Language (UML) – Version 1.5*. OMG document formal/2003-03-01.

[21] Pimentel, A.D., Hertzberger, L.O., Lieverse, P., van der Wolf, P., and Deprettere, E.F. (2001). Exploring embedded-systems architectures with Artemis. *Computer*, 34(11):57–63.

[22] Segala, R. (1995). *Modeling and Verification of Randomized Distributed Real-Time Systems*. Ph.D. thesis, Massachusetts Institute of Technology.

[23] Theelen, B.D. (2004). *Performance Modelling for System-Level Design*. Ph.D. thesis, Eindhoven University of Technology.

[24] van den Bosch, P., Florescu, O., and Verhoef, M. (2006). *Boderc: Model-Based Design of High-Tech Systems*, chapter Modelling of Performance, pp. 103–116.

[25] van der Putten, P.H.A. and Voeten, J.P.M. (1997). *Specification of Reactive Hardware/Software Systems*. Ph.D. thesis, Eindhoven University of Technology.

[26] Wandeler, E., Thiele, L., Verhoef, M., and Lieverse, P. (2006). System architecture evaluation using modular performance analysis: a case study. *Int. Journal of Software Tools for Technology Transfer*.